ISRAEL
AND
CHINA

FROM SILK ROAD
TO INNOVATION HIGHWAY

ISRAEL
AND
CHINA

FROM SILK ROAD
TO INNOVATION HIGHWAY

LIONEL FRIEDFELD   **PHILIPPE METOUDI**

UNDERSTANDING ASIA'S RELATIONSHIP
WITH THE JEWISH STATE

PARTRIDGE
A Penguin Random House Company

ISBN: Hardcover   978-1-4828-5162-5
            Softcover   978-1-4828-5161-8
            eBook   978-1-4828-5160-1

Print information available on the last page.

To order additional copies of this book, contact
Partridge India
000 800 10062 62
orders.india@partridgepublishing.com

www.partridgepublishing.com/india

ENDORSEMENTS

"An enlightening journey into the heart of the Israel–China relationship."

Song Yuanfang, *Professor of Economics, Deputy Dean of Graduate School, People's University of China*

"This book documents the history of gains from trade opportunities between Israel and Asian economies starting in the first millennium, which again have intensified with the innovation-seeking of today's global free markets. Friedfeld and Metoudi offer a compelling demonstration of how free trade of goods and ideas enhances economic development."

Zvi Eckstein, *Dean, Arison School of Business and Tiomkin School of Economics, IDC Herzliya; Former Deputy Governor, Bank of Israel*

"Asia as whole and China in particular will play an important role in Israel's future economic development. This book makes positive contributions for those who are eager to develop and enhance Israel–China relations. Learning and knowing the past is essential for building the future! I congratulate the authors on adding a new perspective to understanding the complexity of relations between Israel and Asia in general and Israel and China in particular."

Dan Catarivas, *First Economic Counselor, Israeli Embassy Beijing; Director, Manufacturers' Association of Israel*

"Lost amid the headlines, China and Israel have established remarkably close economic ties in recent years. This book provides a detailed summary and analysis of the extent and nature of that relationship. An indispensable reference."

Mike Chinoy, *Senior Fellow, USC U.S.–China Institute, former CNN Beijing Bureau Chief and Senior Asia Correspondent*

"A neat and tidy overview, especially for those of us who are novices in the complexities of Israeli–Asian economic relations."

Jacques Berlinerblau, *Professor and Director of the Program for Jewish Civilization at the Edmund A. Walsh School of Foreign Service, Georgetown University*

"This remarkable book gives an overview of the historical connection, dating back over 3,000 years, between Israel, the Jewish people and the Asian continent. Undoubtedly, it will reinforce all the cultural, economic and scientific relationships in our present and our future".

Dr David Harari, *Israel Prize Laureate, "Founding Father" of the Israeli unmanned aerial vehicle industry, former General Manager of Israel Aerospace Industries in Europe*

"For anyone interested in the economic success story which is Israel today, this is a must-read book. It reveals the little known but fascinating historical role of Jews in Asia, and moves on to an in-depth look at Israel's current and growing involvement in the expanding economies of Asia. I highly recommend it."

Barry Topf, *former member of Bank of Israel Monetary Policy Committee and Senior Advisor to former Governor Stanley Fischer, Consultant for International Monetary Fund*

"Israel is a singularity in the tech world. The sheer number of startups, patents and research achievements is stunning – as a consequence, most leading US tech companies have established successful R&D centers in Israel. With the incredible rise of Asia's economy, now is clearly the time for Asian and Chinese businesses to join forces with the Israeli tech ecosystem to achieve greater innovation and business impact. This book is a must-read for anyone interested in considering this opportunity."

Yoelle Maarek, *PhD, VP Research, Yahoo Labs, ACM Fellow, Member of the Technion, Israel Institute of Technology, Board of Governors*

"Anyone looking to understand the burgeoning links between Israel and China should start with this book. Metoudi and Friedfeld offer a convincing explanation of why this relationship has a solid future."

Richard Schwartz, *Director, East3rd Ltd, veteran journalist and commentator on global financial market infrastructure*

"Two historic societies, cultures and economies in so many ways diametrically opposed and in so many ways inextricably linked. It is exactly these anomalies and commonalities that fuel the great opportunities provided by these two economic and industrial systems. The book is an indispensable tool in understanding the great opportunities that have and will continue to grow out of this special relationship."

Mitchell Silk, *Partner and Head of U.S.–China group, Allen & Overy; lecturer at Beijing University, Shenzhen University, Shanghai Institute of Foreign Trade*

FOREWORD

China's economy has been growing at an unprecedented pace in the past few decades and has become one of the driving forces in today's world economy.

To further enhance its international expansion and ability to compete with other leading nations, China has put innovation at the center of its development strategy, sharpening its competitive advantage to an all-time high.

Israel excels at innovation, and the country has gained recognition as one of the world's leading technological powerhouses.

China and Israel have long-standing historical and cultural ties – borne initially from successful interchanges along the Silk Road – that nourished a successful business alchemy. Today, China and Israel come together on the modern stage of globalization and high technology – the Innovation Highway – with complementary strengths that could benefit both countries' strategic initiatives. Israel brings its unique research and innovation capabilities to China, and China offers Israel its financial strength and unparalleled operational capabilities.

Israel and China: From Silk Road to Innovation Highway is an enlightening journey into the heart of the Israel–China relationship.

Song Yuanfang, *Professor of Economics, Deputy Dean of Graduate School, People's University of China*

AUTHORS' NOTE

When we met each other in 1999, more than 15 years ago, at a Friday night function at the Ohel Leah Synagogue in Hong Kong, who could have known that more a decade later these two bankers specialized in corporate finance and asset management would write a book in the summer of 2015 about the new business partnership between Israel and China.

Who could have predicted that Israel would become an innovation and technology global powerhouse?

Who could have foreseen that China would become so powerful in so little time?

We called the book *Israel and China: From Silk Road to Innovation Highway* because it was obvious to us that Israel's and China's partnership in the business arena did not come out of the blue, but rather had a history and deeper meaning that we have tried to shed light on.

History, culture, and spirituality are several of the factors that have played important roles in this successful relationship. Its continuity today is metaphorically symbolized by the "upgrade" of the one of China's strongest historical symbols, the Silk Road, to embrace Israel's strength in innovative high-technology, the Information Highway.

Because we have built much of our lives and careers while in Asia, its history, legends, and success stories have penetrated our lifestyle. At the same time, as Jews, we remain deeply rooted to the land of Israel. We hope that we have managed to describe accurately how Israel and China have come together over the centuries in a unique way based on the key attributes of business, tradition, and innovation.

During our journey we were helped and supported immensely by family members, friends, and advisors who took the time to read, correct, and amend our manuscript and share their opinions with us. We thank them all sincerely.

We hope you enjoy reading the book as much as we enjoyed writing it.

Lionel and Philippe
Tel Aviv, Hong Kong
July 2015

CONTENTS

PART TWO
ISRAEL WORLD INNOVATION CENTER

PART THREE
ASIA WORLD BUSINESS CENTER

INTRODUCTION

"If you want to know where you are going, you must know where you have come from." Pronounced every year by Jewish people all over the world during the festival of Passover, which commemorates the liberation of the Jews from slavery in ancient Egypt, this statement explains the importance of history as a compass to understanding the present and the future. This compass also guides the profound relationship that binds Israel with the countries of Asia.

Israel is often mistakenly believed to be located in the Middle East. In fact it is situated geographically in the Asian continent – precisely in Western Asia on the southeastern shore of the Mediterranean Sea. A core part of Asia with a strategic location, it has long enjoyed a preferential relationship with Asian countries. As the Jewish people were forced into exile with the fall of the Kingdom of Israel and Judah, and following the destruction of the temples of Jerusalem in 586 BCE by Babylonian King Nebuchadnezzar and in 70 CE by Roman Emperor Titus, they emigrated mainly to Babylon (Mesopotamia, present-day Iraq) but also to the Mediterranean basin, Central Asia and the countries of the Indus Valley. There, they continued their activities as traders, buying and selling merchandise between countries.

The creation of the Silk Road during China's Han Dynasty (206 BCE-220 CE) had a tremendous impact on international trade and communication between eastern and western countries. During this golden age, various merchandise, including silk, spices, jewelry, and perfumes, was exchanged on the Silk Road. Having begun to settle alongside this mercantile set of roads, Jewish traders were at the forefront of such activity. Later, the expansion of the European Colonial Empire from the sixteenth to the twentieth century with

Spain (in the Philippines), Portugal (in Japan, Indian ports and Macau), the Netherlands (in Indonesia), Great Britain (in India, Hong Kong–China, Singapore and Burma) and France (in Indochina), led to the rapid growth of international trade. This expansive colonization, more than any other factor, accelerated the dispersion of Jewish communities throughout Asia.

The growing Jewish population across Asia integrated well with its local hosts, establishing large, successful business groups and forming a key component of local cultural and political life. The Kadoorie family of Hong Kong, for example, came originally from Baghdad. They established themselves in India in the mid-eighteenth century, and later in Hong Kong. The Peninsula Hotels Group and CLP Group they founded remain among Asia's largest luxury hotel and electric power generation conglomerates. Individual members of the community, such as David Marshall, who in 1955 became Singapore's first Chief Minister, have made their mark in the political sphere. The Jewish population has also made significant contributions to Asia's infrastructure development through philanthropic charities. The Sassoon family of India, also of Iraqi descent, notably financed two of Bombay's historical landmarks, the Sassoon Docks, built in 1875, and the David Sassoon Library, also constructed in the mid-1800s.

With the establishment of the State of Israel in 1948, the rich common history of the Jewish and indigenous populations in Asia laid the foundation for friendship and cooperation between Israel and many of its fellow Asian countries. The relationship has evolved from the strong ties of the past, symbolized by the Silk Road, to the current strategic partnerships in technology, which we call the Innovation Highway.

In this book, we look at strategic relationships between Israel and the countries of Asia, primarily based on Israel's expertise in innovation and Asia's global position as a center of business, but supported by deep historical, cultural and spiritual links. We outline landmark business transactions and economic factors that have enabled Israel to appear on the roadmap of Asian investors and highlight the activity of some of the major Asian investors in Israel (predominantly from China, India, Singapore, South Korea, Japan and

Taiwan). We also explore how Israel participates in Asian businesses. Finally, we take out our crystal ball and attempt to project the future of Israeli–Asian relationships over the coming fifty years.

Thank you for joining us on this fascinating journey.

PART ONE

OVERVIEW OF JEWISH HISTORICAL PRESENCE IN
ASIA AND RELATIONSHIPS OF THE STATE OF ISRAEL
AND ASIAN COUNTRIES SINCE THE 1950s

CHAPTER 1

Historical Presence of Jewish Population in Asia

Jewish Population of China

During an official visit to China, a high-ranking Chinese official once asked the Chief Rabbi of the Commonwealth how many Jews there are in the world. The rabbi gave a classic and telling reply: "The number," he said, "is smaller than the statistical error of the Chinese census!"

Nevertheless, China and its Jewish population have a rich common history. A Jewish presence was recorded in China as early as the second century. Wayfaring Jewish traders bought and sold merchandise back and forth between China and foreign countries, while some settled more permanently along the Silk Road. Broadly dispersed throughout Chinese territory, from Beijing, Kaifeng, Shanghai, Harbin, Tianjin, Hailar, Hengdaohezi, Mukden, and Qiqihar to Hong Kong, the Jewish population in China followed the country's political and economic climate.

Ancestral Presence of Jews in China

The exile of the Jews into a widespread diaspora began as early as the second century. The majority of Jews relocated to Babylon (today Iraq) but they also spread widely over the Mediterranean basins, Central Asia and the Indus valley. During this time and under the era of Emperor Ming of Han (28-75 CE), many Jews also immigrated to China from Persia. This nurtured a common belief that descendants of the Jewish population in China were part of the ancient lost tribes of the Kingdom of Israel and Judah.

Jewish traders and merchants enjoyed thriving commercial relationships with China. The development of the Silk Road under the Han Dynasty (206 BCE-220 CE) played a central role in reinforcing the bonds between the Jewish and Chinese people. The Silk Road, a large, interconnected roadway established as a commerce route, originally enabled China to sell silk to western countries. Over time, markets for other goods were developed. This network of trade routes became the catalyst and symbol of economic, political, and cultural exchanges between East and West.

The maps in Appendix 1 (pages 171-173) display the Silk Road terrestrial and maritime routes; the ancient kingdoms of David and Solomon, and Israel and Judah; and the migration patterns of the exiled Jews.

An increasing number of Jewish traders and merchants began to settle permanently in cities along the Silk Road and establish residences in China. Numerous historical sources, including books, letters, and administrative documents, record the early presence and activity of Jews in China. A business letter dated 718 CE, written on paper[1] in Judeo-Persian and now in the British Museum's collection, was discovered in 1901 in Danfan Uiliq[2] (Northwest China) by Hungarian British Archeologist Sir Marc Aurel Stein (1862-1943). In 870 CE, Persian author Ibn Khordadbeh (820-912 CE), in his *Book of Roads and Kingdoms* mentioned the presence of Radhanites[3], Jewish merchants who used the Silk Road to reach China. Radhanites are believed to have contributed to the diffusion of the Chinese art of paper-making to the western world. The later annals of the Yuan Dynasty, dated 1329 and 1354, refer to the growing presence of a Jewish population in Beijing, the capital of China. The Venetian explorer Marco Polo (1254-1324) confirmed the sizeable number of Jewish merchants conducting business in Beijing in his book *Marvels of the World*.

---

[1]   At that time paper was manufactured only in China.

[2]   An important post along the Silk Road.

[3]   Medieval Jewish traders who operated businesses along the main commercial roads in Europe, North Africa, and Asia.

Ancient Chinese Jew

Jewish Population of Kaifeng

A few hundred years later, Matteo Ricci (1552-1610), an Italian Jesuit priest based in Beijing, was the first to reveal to Europeans the presence of a settled Jewish population in China, when he first encountered a Kaifeng Jew. The city of Kaifeng, located in mid-eastern China, now part of Henan Province, was home to one of the earliest and largest Jewish populations in the country. Official records attest to a Jewish presence as early as the Tang Dynasty (618-907 CE), and their numbers grew under the Song (960 CE-1127 CE) and Ming (1368-1644 CE) dynasties.

As the capital of the Northern Song Dynasty, Kaifeng was at that time considered a major trade center along the Silk Road. In the seventh century, the city was connected to the Grand Canal, a vast waterway which formed the backbone of China's internal communications and transport system, which enabled trade and commerce to flourish further.

The Jewish population prospered and built a synagogue in 1163. Records of the community were found on three stone slabs, or steles, dated 1489, 1512 and 1663. The inscription on these steles revealed several important facts, including that the Jewish population came to China from India during the Han Dynasty; that seventy Jewish families were given Chinese surnames; that the Jews commemorated the building and rebuilding of the Kaifeng synagogue; that they had an audience with a Song Dynasty Emperor; and that Jewish soldiers in the Chinese Army were said to be "boundlessly loyal to the country".

Kaifeng reached its economic and political golden age during the eleventh century. The city then entered a decline, brought on by severe floods, due to its proximity to the Yellow River. The population declined further, when the Jin-Song wars led Emperor Gaozong to flee south and establish a new capital in the eastern city of Hangzhou. Much of the loyal Kaifeng Jewish community followed the Emperor south.

The Jews who remained in Kaifeng managed to survive and maintain some of their Jewish traditions over the centuries, despite their isolation from other Jewish communities. Priests, scholars, businessmen, and tourists visited what remained of the Kaifeng Jewish community during the eighteenth, nineteenth and twentieth century. With the destruction of the last Kaifeng synagogue in 1860 from floods and fire, however, the community disappeared, leaving only a handful of remaining Kaifeng Jews as living witnesses to its rich history.

Jews of Kaifeng, China

Earth Market Street, Kaifeng (1910),
the location of the Kaifeng synagogue

Jewish Population of Harbin

The city of Harbin (in northern Manchuria, northeastern China) witnessed the influx of a major Jewish presence beginning in the late nineteenth century, following the Russian invasion of Manchuria in 1894-1895. Russia's construction of the Chinese Eastern Railway, an extension of its Trans-Siberian line across Manchuria, offered Russian Jews a way out of the poverty and anti-Semitism they faced in Czarist Russia. The Czar offered the Russian Jews the opportunity to live in Manchuria without restrictions as a way to establish Russia's foothold in China. Jews settled within the city of Harbin and in outlying Manchurian villages.

The railroad brought prosperity to this frontier area as well as the need for goods and services. Harbin's Jewish immigrants helped fill the needs of boom-town Harbin by starting up shops, bakeries, restaurants, newspapers, banks, and heavy industries. Attracted by the chance to live freely in China and earn a living, waves of Russian Jewish refugees continued to settle in the Harbin area throughout the early part of the twentieth century. In the 1920s, the Jewish community's population peaked at about 20,000.

The family of former Israeli Prime Minister Ehud Olmert were among the Russian-born Jewish immigrants fleeing Russia to relocate to Harbin. In 2004, Ehud Olmert came to Harbin to visit the Jewish cemetery and tomb of his grandfather, Joseph Olmert, who died there in 1941.

The Jewish community of Harbin played a significant role in the political, economic, and cultural life of the city. Zionism, which was strictly forbidden in Russia, prospered under the Chinese Harbin influence and the leadership of Abraham Kaufman, who headed the Jewish community for many years. Vestiges of several of Harbin's many Jewish institutions can be seen in the city until today. Harbin's two synagogues have been refurbished, and with the support of the Chinese government, the "new synagogue" currently features the Harbin Jewish History and Culture exhibition.

While the Jewish community represented only 10% of Harbin's foreign population, it controlled 50% of its business trade. Jewish businessman set up major companies such as Harbin Breweries and Harbin Chemicals, and more than thirty leadings firms were owned by Russian Jews. Two of the leading banks located in Harbin were the Harbin Jewish National Bank and the Far East Commercial Bank. Harbin's historical Hotel Moderne, founded by Russian-born Joseph Kaspe, housed a restaurant, cinema and many other amenities, and was a center of Harbin's cultural life.

By the end of World War II, Harbin's Jewish population had diminished significantly on the back of the Japanese army's invasion of Manchuria. Many who were unable to flee the area were repatriated to Russia by the Soviet army. Others relocated to Israel following World War II.

Jewish Population of Shanghai

Shanghai's Jewish population increased dramatically in the second part of the nineteenth century, following the First Opium War, with the arrival of Jewish traders of Middle Eastern origin who came from India under the protection of Great Britain. In 1850, Elias David Sassoon was the first to open a branch of his father's company, David Sassoon & Co, in Shanghai. This leading trading company dealt in numerous products, including cotton yarns and, especially, opium – one of the leading commodities at the time – imported by David Sassoon & Co to China.

The prosperous economic climate encouraged several members of the Jewish community of India to settle in Shanghai. In 1880, a 15-year-old youth, now known as Sir Ellis Kadoorie, joined the Sassoon Group of companies in Shanghai. Within several years he, together with other members of his family, had created an industrial empire of his own with investments spanning rubber plantations, real estate, banking and utilities. A philanthropist as well as a businessman, he was knighted in 1917.

Shanghai's growing economic prosperity created a strong incentive for further Jewish immigration. The Ohel Rachel Synagogue[4], built in the 1920s by Sir Jacob Elias Sassoon, stands as a magnificent example of the golden age of Shanghai's Jewish community, which, at the time, numbered 30,000-40,000 individuals. This buoyant community was considered a key factor in enabling successful development of China's trade activity.

Two historic events in the twentieth century reinforced the numerical strength of Shanghai's Jewish population. In the 1930s, with the Japanese occupation of Manchuria, or northeast China, many Russian Jews who had lived in and around Harbin (in present-day Heilongjiang Province) relocated to prosperous Shanghai. At the same time, the rise of Adolf Hitler's Nazi regime in Germany led more than 20,000 Jews to flee Germany, Austria, Poland, and other parts of Europe and seek refuge in Shanghai. At that time, the city was considered a safe haven for the Jewish population as no visa was required to enter the territory.

At the end of World War II, most of the Jewish population of Shanghai emigrated to Australia, Great Britain, Canada, India, South Africa, Israel, and Hong Kong. In 1998, however, under the leadership of the Chabad Lubavitch movement[5], Rabbi Shalom Greenberg settled in the city. Under his stewardship and with the help of numerous organizations, Jewish life in Shanghai thrives once more.

---

[4] Ohel Rachel Synagogue is a government-protected architectural landmark. Although not in regular use, it is opened occasionally for religious services and special events.

[5] Chabad Lubavitch is the world's largest Orthodox Jewish movement. It strengthened in the second part of the twentieth century under the leadership of Rabbi Menachem Mendel Schneerson. Chabad's global center, based in New York, caters to the needs of the world's Jewish population under the leadership of Rabbis Moshe and Mendy Kotlarsky. Its extensive Asian network, based in Hong Kong and led by Rabbi Mordechai Avtzon, supports Jewish communities and travelers throughout the region.

Mir Yeshiva, Shanghai (1941), was an institute for
the study of traditional Jewish religious texts

Ohel Rachel Synagogue, Shanghai

Jewish Population of Hong Kong

Following the end of the First Opium War (1839-1842) and the signing of the Nanking Treaty that transferred Hong Kong's sovereignty from China to Great Britain, Jewish traders who had previously established offices in Canton (Guangdong Province, China) and Macau (Portuguese settlement, now a Special Administrative Region of China) transferred their activity to Hong Kong to develop this new port. The first Hong Kong Jewish community was formed in 1857, and in 1870 its members opened a synagogue on Hollywood Road. This was eventually replaced in 1901 by the magnificent Ohel Leah Synagogue[6] on Robinson Road, built by Sir Jacob Sassoon in memory of his mother, Leah. In 1904, the Kadoorie family established the Jewish Club.

The Sassoon and Kadoorie families, often described as the "Rothschilds of the East," have stood at the helm of this community. Arthur Sassoon was an original member of the Board of Directors of the Hongkong and Shanghai Banking Corporation (HSBC Holdings). Sir Ellis Kadoorie invested in The Hongkong and Shanghai Hotels, which owns the luxury Peninsula Hotels brand, and in China Light and Power Company (CLP Group), Hong Kong's largest power utility.

By 1954, Hong Kong's Jewish population had grown from around 60 people (mostly Sephardic) in 1860 to 250 (half Sephardic and half Ashkenazi). The relatively small size of the community before World War II can be explained mainly by the attraction of Jewish traders to Shanghai. However, after the 1960s, Hong Kong's development as a global center of trade and finance attracted thousands of Jewish expatriates of all nationalities. A support infrastructure of outstanding facilities has grown up around this burgeoning community of more than 5,000. These include the Jewish Community Center; Carmel School; Elsa High School; seven synagogues; Chabad Lubavitch's Asian headquarters; kosher restaurants; and a Jewish library.

---

[6]   Ohel Leah Synagogue, led by Rabbi Asher Oser, functions today as a center of Hong Kong's Jewish community. The synagogue underwent a full restoration in 1996-1998, and received an *Outstanding Project Award* in the inaugural *UNESCO Asia-Pacific Heritage 2000 Awards*.

Ohel Leah Synagogue, Hong Kong (circa 1920)

Jewish Population of India

India historically has hosted one of the world's oldest and most diverse Jewish communities, including the ancestral community of Cochin, the historical tribe of Bene Israel, and the business community of Baghdadis. India's rich Jewish history embraces many other lesser-known tribes of Jewish origin who followed Judaic rites, such as the Bene Menashe and Bene Ephraim. The majority of India's Jewish community, at one time numbering about 70,000, emigrated to Israel following the creation of the State of Israel in 1948.

Cochin Jews

It is believed that India's first Jewish residents, mostly traders from the Kingdom of Judea, arrived in Cochin, southern India (Kerala state) following the destruction of the first temple of Jerusalem by Babylonian King Nebuchadnezzar in 586 BCE. The flow of Jewish arrivals quickened in 70 CE when the second temple of Jerusalem was destroyed by Roman Emperor Titus. They established themselves in Cranganore, an ancient port near Cochin. The Indian rulers, the

Ceras of Cranganore, gave Jewish merchant Joseph Rabban the title of "Prince of the Cochin Jews". At the same time, they granted him property and tax-levying rights, which were outlined on a copper plate dated around 1000 CE. The first Cochin Jews were known as Malabar Jews or "Black Jews". They built a synagogue in 1200 and enjoyed prosperity and growth throughout the years, working mostly as traders and merchants specialized in commodities and spices, particularly pepper.

The text on the copper plate from India (1000 CE)
granting land to Jewish trader Joseph Rabban

Following the expulsion of Jews from Spain in 1492 and from Portugal in 1496, another group immigrated to Cochin. This community was known as Paradesi Jews or "White Jews". Their language skills enabled them to manage and maintain prosperous trade connections with Europe, while learning Judeo-Malayalam from the Malabar Jews. In 1568, the Synagogue Paradesi was built in Cochin by these traders. In 1968, the State of India issued a postal stamp celebrating the 400-year anniversary of this landmark building, which is today the oldest active synagogue in the Commonwealth.

Group of Cochin, India Jews (circa 1900)

Inscription on the oldest synagogue in Cochin, built in 1344

Bene Israel

Considered one of the most ancient Jewish communities in India, the Bene Israel believe they belong to the lost tribes of the Kingdom of Judea. They came to India after centuries of wandering through Asian countries, settling originally in the Konkan area along the west coast of India, before relocating to Bombay, Pune, Calcutta, Ahmedabad, and Karachi. At its peak, this community amounted to some 20,000 individuals, some of whom were particularly active in the British Colonial Administration and in the nascent film industry as directors, producers, and actors. Mumbai had a dynamic Jewish population until the 1960s when the bulk of the population emigrated to Israel. A small, but active, community remains.

Baghdadi Jews

The first Jew from Baghdad, Joseph Semah, arrived in the city of Surat (Gujarat state) in 1730. Other Baghdadi Jews followed and settled in Bombay. There, they established a synagogue and other community facilities. The Jewish population in Bombay was mainly composed of successful merchants and traders. The Sassoon family was one of the most predominant Baghdadi families; its patriarch, David Sassoon, left Baghdad in 1832 to settle in Bombay. He then sent his children all over Asia (Singapore, Macau, Burma, Canton, Japan, Shanghai, and Hong Kong) to open branches of David Sassoon & Co. The Sassoon family made significant contributions to the construction of India's infrastructure and financed important philanthropic projects, including the landmark Sassoon Docks and David Sassoon Library.

Synagogue Eliyahu, Mumbai (circa 1900)

Jewish Population of Singapore

In 1819, Sir Stamford Raffles, on behalf of the British East India Company, signed an agreement with Sultan Hussein Shah of Johor to establish a British trading post and port in Singapore. The well-located port captured the attention of Baghdadi Jewish traders from India who rapidly made Singapore their new home. The Indian-based Sassoon family had opened a representative office there in 1840. The following year, the first synagogue, Maghain Aboth, was erected on Synagogue Street. In 1878, as the Jewish population in Singapore grew, a new and larger Maghain Aboth Synagogue[7] was inaugurated on Waterloo Street.

The Singapore Jewish community was led for many years by Menashe Meyer, a Baghdadi Jew from India who arrived in Singapore at the age of 18. By 1873, he had successfully established an import/export company, eventually becoming Singapore's wealthiest person. Meyer, who was later knighted by the Queen of England, was said to own half of Singapore's real estate, including landmark hotels, and was a leading opium[8] trader. In 1905, Sir Menashe Meyer built a second synagogue, Chesed-El, on the grounds of his luxurious residence on Oxley Rise. By the 1930s, Singapore's Jewish population had grown to around 830.

Another important figure in the Singapore Jewish community, David Marshall (1908-1995), was born in Singapore to a Baghdadi Indian family and served as Singapore's first Chief Minister from 1955 to 1956. Marshall also founded one of Singapore's two leading political parties, the Workers' Party of Singapore. From 1978 to 1993, he served as Ambassador of Singapore to France, Portugal, Spain, and Switzerland. Marshall also served for many years as the President of the Singapore Jewish community.

Today, the Jewish population of Singapore numbers about 300, comprised of Singaporeans of Baghdadi and Indian origins, as well as many expatriates.

---

[7]   Religious life at the Maghain Aboth and Chesed-El Synagogues remains active, with daily services, adult education, and other community activities.

[8]   This commodity was legal under British rule.

Jewish Population of Japan

The signing of the Treaty of Kanagawa[9] opened the doors of international trade with Japan. In 1861, the first Jews (the Marks brothers) arrived in Yokohama. Raphael Schover, an American merchant, followed in their path and later established Japan's first foreign newspaper, *Japan Express*. Jews of Polish and British origin also settled in the city, and in 1895 Yokohama's Jewish community of approximately 50 families inaugurated its first synagogue. The Great Kanto Earthquake of 1923 forced the community to relocate to Kobe.

In 1880, a wave of Russian and other Jewish immigrants settled in Nagasaki, which had been a free port since 1859 and a leading port for international commerce. In 1894, the Nagasaki community of more than 100 Jewish families inaugurated the Beth Israel Synagogue. Joseph Trumpeldor, a leading member of this community, eventually became a leading figure in the Zionist movement advocating the establishment of a Jewish state in the Jews' historic homeland of Israel, and one of the founders of the Jewish Defense Forces – the precursor of today's Israel Defense Forces. In 1923, following the Great Kanto Earthquake, this community also moved to Kobe.

Beth Israel Synagogue, Nagasaki, Japan (circa 1900)

---

9   This treaty, signed in 1854 between Commodore Matthew Perry (United States) and Tokugawa Shogunate (Japan), opened Japanese ports to the United States Navy.

In the 1900s, many more Russian Jews immigrated to Japan following the Russian Revolution of 1905 and the Bolshevik Revolution of 1917. They settled mostly in Tokyo, Yokohama, and Kobe. A synagogue was established in Kobe in 1937 and from the 1930s until the 1950s, the Jewish population of Kobe was the largest in Japan.

In 1953, Tokyo's first synagogue was inaugurated. Since then, the Jewish population has been composed mainly of international expatriates. It is well served by diversified and dynamic Jewish institutions. Today, several hundred Jewish families reside in Japan, mostly in Tokyo.

Jewish Population of South Korea

The first Jewish population arrived in South Korea as part of the United States military during the Korean War (1950-1953). Author Chaim Potok (1929-2002), a United States Army Chaplain at the time, wrote *The Book of Lights* following his Korean experience. Today, most of Korea's Jewish community resides in Seoul. It is composed of expatriates, who are well served by various institutions.

Judaism has fascinated South Koreans for many years, both because of its long, rich history and its strong ethical values. Today, many South Koreans keep a translated copy of the Talmud[10] in their home.

Jewish Population of Taiwan

The Jewish presence in Taiwan is recent, dating from the arrival of the United States military in the 1950s. In 1975, Rabbi Ephraim Einhorn was appointed as Chief Rabbi of Taiwan to cater to the needs of the growing Jewish business community. Today, American, French and Israeli expatriates dominate Taiwan's Jewish community, which numbers about 400. Activities are held in several hotel facilities.

---

[10]   The Talmud is one of Judaism's most important Rabbinical books; it contains the opinions and teachings of numerous renowned Rabbis over a variety of subject matters, including law, ethics, and philosophy.

Jewish Population of Burma (Myanmar)

The first record of a Jewish presence in Burma dates from the eighteenth century when Salomon Gabirol served in the army of King Alaungpaya. In the nineteenth century, the Jewish population increased dramatically with the settlement of Indian Jews of Cochin and Baghdadi origin, mostly drawn by the trade of opium, cotton, and rice. In 1852, Great Britain conquered the capital Rangoon and the flow of Jewish traders accelerated. In 1857, the first synagogue, Musmeah Yeshua[11], was built. In the 1890s, Jewish merchants of diverse origins began arriving in Rangoon. These included Jonas Goldenberg from Romania, a wealthy teakwood merchant, and Solomon Reineman from Galicia, Central Europe who became a supplier to the British Army.

By 1901, the Jewish population numbered more than 500. In 1932, a second synagogue, Beth-El, was constructed. The Jewish community's influence expanded in the 1930s and 1940s. Rangoon appointed a Jewish mayor, David Sofaer, and the Jewish population reached its peak of 2,500. Since then, the population has declined considerably with most Jews emigrating to Israel. Notably, Burma, which – like Israel – achieved its independence from Britain in 1948, was one of the first Asian nations to grant the Jewish state formal diplomatic recognition.

Jewish merchants in front of Sofaer & Co
building, Yangon, Myanmar

---

[11]   Musmeah Yeshua Synagogue is well preserved and managed by a caretaker. Although regular services no longer take place, the synagogue remains open to visitors.

Jewish Population of Indonesia

Jews arrived in Indonesia in the seventeenth century as traders and employees of the Dutch East India Company. They came from the Netherlands and settled in Jakarta, Surabaya, and Semarang (Java island). In the eighteenth century, another influx of Jews of Dutch and German origin settled mostly in Jakarta. Indonesia's largest synagogue, Beth Shalom[12], was inaugurated in Surabaya in the nineteenth century. The country's Jewish population peaked at about 2,000 in the 1920s with the immigration of Baghdadi Jews, mostly traders specialized in spices, to Surabaya. During the same period, Jews from the Netherlands and Aden (Yemen) also settled in various locations. Most of Indonesia's Jewish population emigrated to the United States, Australia, and Israel during the 1940s and 1950s.

Bar mitzvah celebration, the Jewish rite
of passage, Surabaya, Indonesia

---

[12] Beth Shalom Synagogue was designated a heritage site by the Surabaya Tourism Agency in 2009. It was destroyed by vandals in 2013.

Hanukkah celebration, the Jewish Festival of
Lights, Bandung, Indonesia (1935)

Jewish Population of the Philippines

Following the expulsion of the Jews from Spain in 1492, many crypto-Jews, who were forced to convert to Christianity but secretly continued to follow Judaism, tried to reach the new colonies of Spain, including the Philippines. Records show that two brothers of Jewish origin, Jorge and Domingo Rodríguez, arrived in the Philippines in the 1590s and over time a small community of *marranos* formed in the islands.

Later, during the Spanish colonial period, a growing Jewish population settled in the Philippines encouraged by the prospect of successful trade ventures. The Levy brothers, Charles, Adolphe, and Raphael, came from France's Alsace-Loraine region fleeing from the Franco–Prussian War in 1870. They successfully managed a trading company located in Iloilo City, and then in Manila, with activity in jewelry, gemstones, pharmaceuticals, and general merchandising. Leopold Kahn from France, another leader of the community, was president of several companies, Consul General of

France and President of the French Chamber of Commerce. In 1869, the inauguration of the Suez Canal dramatically shortened the trade route between Europe and the Philippines. Over time, strong economic growth attracted Jews from Egypt, Turkey, and Syria to the Philippines and the Jewish population rose to approximately fifty individuals.

In 1898, the United States took control of the Philippines from Spain and Judaism was officially recognized. The arrival of the American military brought Jewish teachers and businessmen to the Philippines. One of the most highly regarded American Jews, Russian-born Emil Bachrach, arrived in Manila in 1901 and created the Bachrach Motor Company, an automobile distributorship and the largest operator of taxis in Manila. Another American family, the Frieder brothers Philip and Alex, arrived in the Philippines in 1918 to expand their father's New York-based S. Frieder & Sons cigar activity. They eventually opened Helena Cigar Factory and sold products under various brands in the United States. Jews held prominent positions in Filipino society, including as members of the Makati Stock Exchange (now the Philippines Stock Exchange), physicians, architects, and a conductor of the Manila Symphony Orchestra. In 1922, Emil Bachrach's wealth and philanthropic activity enabled the construction of Temple Emil Synagogue[13] and Bachrach Memorial Hall in Manila.

Temple Emil Synagogue, Manila (1940)

---

[13] Temple Emil Synagogue is now a commercial building owned by the Jewish community of the Philippines.

The Philippines Jewish population had grown to around 500 by 1936, and during World War II, 1,300 refugees from Europe reinforced the community. Following the difficult period of Japanese occupation, the community peaked at approximately 2,500, and then began to decline. The current Jewish population of several hundred mainly consists of American, French and Israeli expatriates, among others, as well as diplomats.

Community Passover seder, Manila, Philippines (1925)

Jewish Population of Thailand

In 1601, Spanish missionaries reported the presence of Jewish traders in the Kingdom of Ayutthaya in Siam, the former name for Thailand. In 1683, Abraham Navarro, a Jewish translator for the East India Company, visited the court of King Narai in Lopburi. During the 1800s, a growing number of Jewish traders and merchants came and settled in Thailand. In 1890, during the reign of King Chulalongkorn, several European families settled in Bangkok. One of the most prominent, the Rosenberg family, developed Europe Hotel, the first modern hotel in the country. They were joined in the 1920s by Russian Jews fleeing the post-revolution turmoil.

Haim Gerson, a successful businessman, led the Jewish community for several decades. During the 1930s and 1940s, Jews from Germany, Lebanon and Syria settled in Thailand. They were following in the 1950s and 1960s by growing numbers of Jews from the United States, as well as Jews from Iran, Afghanistan, and Iraq who were escaping persecution.

In the 1970s, an additional influx of Jewish American military troops who had been based in Vietnam made Thailand their home. Bangkok's Beth Elisheva Synagogue[14] was inaugurated in 1979. Today, Thailand's Jewish community consists mostly of expatriates and seasonal travelers from around the world.

Jewish Population of Indochina (Vietnam, Laos, Cambodia)

Jews first settled in Saigon, Indochina (now Vietnam, Laos, Cambodia) during the French colonization period (1852-1954). Jules Rueff, considered as one of Indochina's leading Jewish pioneers, arrived from France in 1872 and developed the Cochichine region's railroad system as well as the steamship company Messageries Fluviales de Cochichine. Between 1883 and 1886, Jewish officers fought in the Tonkin Campaign; Captain Louis Paquet was awarded several medals of honor including the French Chevalier of the Order Royal du Cambodge. In 1902, The French School of the Far East, L'Ecole Française d'Extrême-Orient, was created by Jewish intellectual Sylvain Lévi in Hanoi. In the 1920s, the Alliance Israélite Universelle (secular Jewish school) also was active in Haiphong.

By the 1940s, the Jewish population of Indochina (Hanoi, Saigon, and Tourane) had grown to around 1,000. After 1954 and the dissolution of French Indochina, most of the Jewish population returned to France. Today, Vietnam, Laos, and Cambodia have small Jewish communities, mostly comprised of American, French and Israeli expatriates, among others, and seasonal travelers. Chabad Lubavitch is active in these countries.

---

[14]   Beth Elisheva Synagogue, Mikveh and Jewish Center, led by Chabad Lubavitch Rabbi Yosef Kantor, functions until today as the center of Bangkok's Jewish community.

CHAPTER 2

Leading Jewish Individuals in Asia

Sassoon and Kadoorie Families: Rothschilds of the East

The Sassoon and Kadoorie families were among the leading Jewish families in Baghdad (Iraq), together with the Somech, Sopher, and Gubbay families. Their destinies were, however, completely recast, when they reached India. They actively participated in the opening of trade in British outposts in Bombay, Calcutta, Shanghai, Hong Kong, Canton, and Singapore. Their wealth, based on commodities trading, especially opium, was invested in numerous industrial projects such as hotels, real estate, transportation, utilities, and banking. These two families, often related by marriage, became known as the "Rothschilds of the East" because of their financial strength, political power, and philanthropic generosity.

The Sassoon Family

David Sassoon (1792-1864)

David Sassoon was the son of Saleh Sassoon (1750-1830), the Chief Treasurer of the Governor of Baghdad from 1781 to 1817 and president of the Jewish community for many years. Proficient in several languages, including Arabic, Persian, Turkish, Hebrew, and Hindustani, David Sassoon worked in a bank until 1822. Fleeing political instability, he left Baghdad with his family to the city of Basra (Iraq) and in 1828 reached the Gulf port of Bushehr (Iran).

In 1832, Sassoon settled in Bombay and started to work as a middleman for the British East India Company, subsequently founding David Sassoon & Co, which traded in commodities (wheat, spices, rice, sugar, tea), precious metals (silver, gold), garments (silk, cotton yarn), and opium. He took advantage of the Treaty of Nanking in 1842, which opened up China trade to British merchants, to establish branches of his parent company in Calcutta, Hong Kong, Shanghai, and Canton with the help of his sons and employees of Baghdadi origins.

David Sassoon & Co's successful operation was driven by a very profitable triangular business, based on the export of Indian cotton yarn and opium to China, the purchase of Chinese merchandise (tea, silk) and its export to Great Britain, and the purchase of English Lancashire[15] textile products to import back to India. The de facto monopoly of David Sassoon & Co in the trade of opium to China enabled its founder to become the wealthiest person in Bombay.

David Sassoon founded numerous charitable organizations and developed his philanthropic activities throughout Asia. In India, he built gardens, synagogues, schools, orphanages, hospitals, museums, and national monuments. Some of his most notable contributions include Victoria Garden and Albert Museum, the Magen David and Knesset Eliyahu synagogues in Mumbai, Ohel David Synagogue in Pune, David Sassoon Reformatory and Deaf school, David Sassoon and Masina hospitals, Bank of India Fort and others. In 1853, Sassoon was naturalized as a British citizen. Upon his death in 1864 he received tributes from around the world for his many contributions to society.

---

[15]   At one time, Lancashire provided 40% of the world's cotton fabric.

David Sassoon

Sir Albert Abdullah David Sassoon (1818-1896)

Sir Albert Abdullah David Sassoon was the eldest son of David Sassoon. He was educated in India and worked with his father at David Sassoon & Co, developing the firm's trade relationship with China and its broader opium trade. David Sassoon & Co controlled an estimated 70% of the opium traded in Bombay and Calcutta.

Albert Sassoon took over the family business when his father passed away. Famous for his philanthropic activity, he built the Sassoon Docks in Bombay in 1875, the first wet dock on the west coast of India. He also developed his firm's business in Persia. In 1871, in recognition of his contribution to the development of trade in Persia, he was made a member of "The Order of the Lion and Sun" by the Shah of Persia. In 1867, Britain also recognized his leading contributions to public service, awarding him "the Star of India". He later became a member of the Bombay Legislative Council.

In 1872, he was made "Knight of Bath" and, in 1873, was granted "the freedom of the city of London", following which he decided to settle in England. *Vanity Fair* famously portrayed Sir Albert Abdullah

David Sassoon as "The Indian Rothschild". He was made Baronet before passing away in 1896 in Brighton, England.

Sir Elias David Sassoon (1820-1880)

Sir Elias David Sassoon was the second son of David Sassoon. His father sent him to China in 1844 to open branches of David Sassoon & Co in Hong Kong, Shanghai, and Canton. At that time, Chinese cotton had become a major global commodity, as the American Civil War (1861-1865) prevented southern states from supplying cotton to factories. Elias David Sassoon and David Sassoon & Co were among the largest cotton traders in the world.

Sassoon eventually returned to Bombay and managed his father's business. He created his own trading company, E.D Sassoon, in 1867 and established offices in Bombay, Hong Kong, and Shanghai. Over time, he also developed activity in the Gulf ports, Baghdad, and Japan. In keeping with the family tradition, Sir Elias David Sassoon was a great philanthropist. He financed the construction of numerous institutions including the Maternity Hospital and the David Sassoon Infirm Asylum in Pune, India.

David Sassoon (seated) and his sons, Elias David,
Albert Abdullah David, and Sassoon David

Sir Ellice Victor Sassoon (1881-1961)

Sir Ellice Victor Sassoon was the younger son of Sir Edward Elias Sassoon and grandson of Sir Elias David Sassoon. He lived in Shanghai where he perpetuated the family's long tradition of successful business ventures. In 1929, he built Shanghai's first high rise – ten floors and seventy-seven meters high – located on the prestigious Bund promenade. It comprised the Sassoon House and Cathay Hotel (now known as the Peace Hotel), the most luxurious landmark hotel in Shanghai. The Sassoon House hosted Sassoon's companies, subsidiaries, and other offices; the top floor was Sir Ellice Victor Sassoon's private apartment. In 1948, Sir Ellice Victor Sassoon sold his business interests in Shanghai and India and transferred its assets to the Bahamas. In 1952, he created the E.D Sassoon Banking Company, which was eventually purchased in 1972 by the merchant bank Wallace Brothers and Company and acquired in 1976 by Standard Chartered. Like his predecessors, Sir Ellice Victor Sassoon was a philanthropist who financed many charities throughout his lifetime.

The Kadoorie Family

Sir Elly Kadoorie (1867-1944)

Sir Elly Kadoorie was the son of Salih Kadoorie from Baghdad. In the late 1870s, he moved to Bombay to work for E.D Sassoon & Co and in 1880, he joined the Sassoon family's Hong Kong office. He was subsequently sent to Shanghai to supervise the Sassoon's business activity in mainland China (Tianjin, Ningbo, Wuhu, Weihai).

Sir Elly established his own company in 1890. With a number of partners, he launched Hong Kong's first brokerage firm: Benjamin, Kelly & Potts. Later he founded the trading and investment firm E.S Kadoorie & Co, with offices in Shanghai and Hong Kong. The firm invested in rubber plantations, hotels, property, wharves, docks, and gas.

In 1901, he was part of the syndicate that founded China Light and Power company (now called CLP Holding[16]), one of Hong Kong's largest electric utility companies. In 1914, he invested in the Hong Kong and Shanghai Hotels, which operates the landmark Peninsula Hotel in Hong Kong, opened in 1928. The brand eventually expanded across Asia, Europe, and the United States as a luxury hospitality group. Sir Elly was a grand philanthropist; he built numerous schools and hospitals in the Middle East and financed many charities.

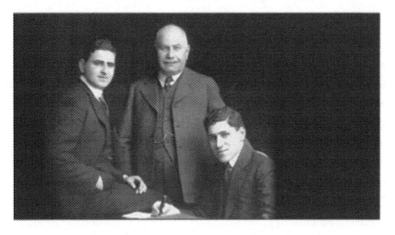

Sir Elly Kadoorie (center) and his sons, Lawrence and Horace

Sir Ellis Kadoorie (1865-1922)

Sir Ellis Kadoorie was the son of Salih Kadoorie from Baghdad and the brother of Sir Elly Kadoorie. In 1883, he left Bombay and joined his brother Elly in Hong Kong to work for the Sassoon family. He and his brother managed the family interests in real estate and properties, rubber, hotels, and utilities. Sir Ellis Kadoorie was the Director of The Hongkong and Shanghai Hotels.

Sir Ellis was knighted in 1917. He was well known for his philanthropic activities, notably as a founder of schools in China,

---

[16]   CLP Holding is the second-oldest stock listed on the Hong Kong Stock Exchange with ticker no. 2 after Li Ka-shing's Cheung Kong Holdings with ticker no. 1.

including the Yucai High School in Shanghai and the Sir Ellis Kadoorie Secondary School in West Kowloon in Hong Kong. Following instructions in his will to finance education in the Holy Land, two schools were built upon his death: the Kadoorie Agriculture High School[17] close to Mount Tabor in what is now northeastern Israel, and the Kadoorie Institute in Tulkarem in the present-day West Bank. He also financed numerous schools and institutions in India, Iraq, Iran, Syria, Turkey, France, Portugal, England, and China.

Sir Lawrence Kadoorie (1899-1993)

Sir Lawrence Kadoorie, born in Hong Kong, was the son of Sir Elly and the brother of Sir Horace Kadoorie. During the 1920s and 1930s, he worked for Victor Sassoon, managing the landmark Cathay Hotel in Shanghai. He later joined the family business and, together with his brother Horace, managed their interests in rubber, cotton, hotels, docks, utilities and real estate, including the Kadoorie estate's eight hectares of houses and low-rise buildings in Kowloon.

Following World War II, the reconstruction of the ruined CLP power plant provided electricity for Kowloon and the New Territories and contributed greatly to Hong Kong's strong economic recovery. In the 1950s, Sir Lawrence Kadoorie also established Tai Ping Carpets in Hong Kong – today the world's largest manufacturer of hand-made tufted carpets. From 1951 to 1954, Sir Lawrence served on Hong Kong's Executive Council and also took the reins of the family business. From 1957 to 1967, he served as a member of the Board of Directors of The Hongkong and Shanghai Banking Corporation (HSBC Hong Kong).

Over the years, Sir Lawrence successfully developed new business activities and investments, notably in Hong Kong, with stakes in The Peak Tram, the Cross Harbour Tunnel, and the Star Ferry. In 1974, he was knighted and in 1981 was granted the title of "Baron of Kowloon and the City of Westminster" by Great Britain. Numerous

---

[17]   Yitzhak Rabin, Israel's former Prime Minister and Nobel Prize Laureate, and Yigal Allon, a former General of the Israeli Defense Forces, attended the school.

other countries, including Belgium, the Philippines, and France also honored him with awards and medals, such as French "Chevalier de la Legion d'Honneur".

Sir Lawrence was the President of the Hong Kong Jewish community and an active philanthropist. His son Michael Kadoorie now heads the family. According to *Forbes* 2015 survey, he is Hong Kong's seventh-wealthiest individual and possesses among the largest fortunes in the world with an estimated US$9 billion in assets.

Sir Horace Kadoorie (1902-1995)

Sir Horace Kadoorie, born in London, was the son of Sir Elly Kadoorie and brother of Sir Lawrence. During the 1920s and 1930s, he worked with his brother managing the Cathay Hotel in Shanghai for Victor Sassoon and overseeing the family interests in rubber, cotton, hotels, real estate, docks, and utilities. Sir Horace was for thirty-five years the Chairman of The Hongkong and Shanghai Hotels. In 1949, he became president of the new Jewish Recreation Club in Hong Kong and in 1951, together with his brother, founded the Kadoorie Agricultural Aid Association, providing training in sustainable agriculture and enabling hundreds of thousands of rural area farmers in the New Territories to gain independence. Later, as agriculture declined, the Kadoorie Farm shifted its focus to environmental activities. In 1962, he received the Philippines' "Ramon Magsaysay Award" for public service as well as the "Gorkha Dakshina Bahu" distinction from the Nepalese government for training the country's farmers. Following the family tradition, Sir Horace was a philanthropist, who financed many charitable associations throughout his lifetime.

Selected Portraits of Outstanding Jews in Asia

Over the centuries, there have been numerous outstanding Jewish individuals in Asia. The handful of leading figures profiled below represent the spirit of excellence and entrepreneurship that categorized the Jewish population in Asia. A prince, political leader, civil servant, trader, and beauty pageant winner join the businessmen in our portrait gallery.

China and Hong Kong

Zhao Yingcheng (1619-1657)

Zhao Yingcheng was a preeminent Kaifeng Jew. In 1646, proficient in Chinese and Hebrew, he obtained his *jinshi* imperial degree. He was appointed Director of the Ministry of Justice and was later sent to serve as a representative in the provinces of Fujian (south coast of China) and Huguang (Hubei and Hunan provinces). Zhao Yingcheng was an excellent administrator. In 1642, at the end of the Ming Dynasty, Kaifeng began to decline, hit by numerous floods from the nearby Yellow River and the Jin-Song wars. Kaifeng's Jewish community followed Emperor Gaozong to Hangzhou. A decade later, however, Zhao Yingcheng restored Kaifeng and in 1653 rebuilt the synagogue, with the help of his brother Zhao Yingdou. A stele dated 1663 records his actions.

Zhao Yingcheng
(Hebrew name: Moshe ben Abram)

Emanuel Raphael Belilios (1837-1905)

Emanuel Raphael Belilios was born in Calcutta and settled in Hong Kong in 1862 after his marriage to Simha Ezra. During the 1870s, he was the Chairman of The Hongkong and Shanghai Hotels. From 1876 to 1882 he was named Chairman of The Hongkong and Shanghai Banking Corporation and in 1881 was appointed to Hong Kong's Legislative Council. Belilios contributed to numerous charities and philanthropic activities, including financing Chinese medical students and public schools for girls in Hong Kong.

Emanuel Raphael Belilios

Silas Aaron Hardoon (1851-1931)

Born in Baghdad, Silas Aaron Hardoon moved with his family to India. He studied in Bombay in a school funded by David Sassoon. In 1868, he immigrated to Shanghai, where he worked as an employee for David Sassoon & Co. He advanced quickly and joined E.D Sassoon as a partner specializing in cotton and opium trading and real estate, on which he later focused his activity. He owned numerous properties on Nanking Road, considered the most prestigious avenue in Shanghai. In 1927, he built the Beth Aharon Synagogue in Shanghai. When Silas Aaron Hardoon passed away in 1931, his estimated fortune of US$650 million made him one of the wealthiest men in Asia.

Hardoon was as a philanthropist who funded many charitable organizations. A widely recognized patron of the arts, he was active in numerous cultural associations as well as a publishing house. Hardoon was portrayed in the literature of the time as a traditional Chinese merchant-philanthropist wearing Chinese clothing. He was also the only foreigner ever to appear as a character in a Chinese opera!

Silas Aaron Hardoon

Sir Matthew Nathan (1862-1939)

Sir Matthew Nathan, a British soldier and civil servant, was born in Paddington, England and educated at the Royal Military Academy and the School of Military Engineering in England. He participated in military expeditions in Sudan (1884-1885) and India (1889-1894) and was promoted to Captain in 1899. Sir Matthew was appointed Governor of Sierra Leone (1899-1900), then Governor of the Gold Coast (today Ghana) until 1903.

Named Governor of Hong Kong in 1903, he remained in that position until 1907. Sir Matthew introduced dramatic structural changes in the British colony, such as the establishment of central urban planning, the urbanization of the Kowloon Peninsula, and the construction of the Kowloon–Canton Railway. Hong Kong's Nathan Road, the main commercial road in Kowloon, is named after him.

After a successful tenure in Hong Kong, he was sent by the British Government in 1907 to be Governor of Natal in South Africa. He continued his civil service career as Chairman of the Board of Inland Revenue (1911-1914), Under Secretary for Ireland (1914-1916), Secretary to the Ministry of Pensions (1919), and Governor of Queensland (1920-1925).

Sir Matthew Nathan

Edward Shellim (1869-1928)

Edward Shellim was born in England and was the grandson of David Sassoon. He joined David Sassoon & Co in Hong Kong, working as the branch manager until 1918. Shellim was the Chairman of The Hongkong and Shanghai Banking Corporation from 1908 to 1912. He also served on the Board of Directors of Hong Kong Tramways, Hong Kong Kowloon Wharf & Godown Co, Hong Kong Fire Insurance, Hong Kong Land Investment & Agency Co, Hong Kong Reclamation Co, and China Sugar Refining Co. He was a member of the Hong Kong Chamber of Commerce, Chairman of the Financial Committee of the Alice Memorial Hospital, and President of the Ohel Leah Synagogue. Shellim was a philanthropist who also was very active in Shanghai institutions.

Edward Shellim

Edward Issac Ezra (1882-1921)

Born in Shanghai, Edward Issac Ezra was considered one of the wealthiest individuals in the city. He was also a member of the Shanghai Municipal Council. His diverse business activities included the Central Stores Company, which took over the Central Hotel (formerly Cathay Hotel). He also managed numerous properties on Nanking, Kiujiang, Szechwan and Kiangse roads and owned several other hotels, including the Astor House Hotel on the Bund.

Ezra was the chairman of several companies, including the Far Eastern Insurance Company, the Shanghai Gas Company, China Motors Ltd, and the *China Press* and *Evening Star* newspapers. Active in the opium business, he was elected first President of the Shanghai Opium regulatory body in 1913. Maintaining a wide range of philanthropic activities, he was President of the Shanghai Zionist Association and Vice President of the Jewish Organization of China.

India

Joseph Rabban

Joseph Rabban was a Jewish trader who settled in the Malabar Coast of India (now Kerala state). The Indian rulers, the Ceras of Cranganore (a small port close to Cochin) granted him the title of "Prince of the Cochin Jews". As a result, he attained a series of property and tax rights, which were outlined on a copper plate dated around 1000 CE.

Joseph Azar

A fourteenth-century descendant of Joseph Rabban, Joseph Azar was last in the line of descendants of Joseph Rabban, the Prince of the Cochin Jews. Azar thereby inherited the title of Jewish Prince Anjuvannam of Cochin, and was the last Jewish prince in India. The Jews placed themselves under the protection of the Maharajah of Cochin. In 1340, a succession conflict broke out with his brother that led to the disappearance of Jewish autonomy in the south of India.

Singapore

Sir Manasseh Meyer (1843-1930)

Sir Manasseh Meyer was born in Baghdad and initially educated in Calcutta, India. In 1861, at the age of 18, he traveled to Singapore to continue his education at St Joseph's Institution, now the third-oldest secondary school in the city. In 1864, he returned to Calcutta to learn Hebrew and Arabic, following which, his business interests took him around the region.

In 1867, he went to Rangoon, Burma, where he created a business. Six years later, he returned to Singapore and founded Meyer Brothers, which became Singapore's largest trader of Indian merchandise, including opium. As he accumulated wealth, he successfully invested in real estate with purchases of numerous properties, including Adelphi Hotel, Meyers chambers at Raffles Place, Sea View Hotel, Meyer Mansion, and Killiney House. Sir Manasseh Meyer was said to own half of Singapore's real estate.

From 1893 to 1900 he served as a Municipal Commissioner and was later appointed a member of the Straits Committee on Currency. Sir Manasseh Meyer was a large donor who contributed to numerous charities. He financed Singapore's two synagogues, Maghain Aboth and Chesed-El, and set-up the Sir Manasseh Meyer Trust that funded educational institutions such as the Raffles College (now part of the National University of Singapore). In 1929, he was knighted by Britain for his outstanding contribution to society.

David Saul Marshall (1908-1995)

David Marshall was born in Singapore into a Baghdadi Indian family and studied in local institutions (Saint Joseph's, Saint Andrew's, and Raffles). He read law in London and returned to Singapore to practice criminal law. In 1955, he led the Labour Party to victory, forming a government in which he served as Chief Minister (1955 to 1956). Marshall also founded one of the two leading political parties in Singapore, the Workers' Party of Singapore, and was a member of the Parliament from 1957 to 1963. He played a critical role in the negotiation of Singapore's independence from Great Britain. Between 1978 and 1993, Marshall was appointed Ambassador of Singapore to several countries, including France, Portugal, Spain, and Switzerland. He also served for many years as the President of the Singapore Jewish community.

David Saul Marshall

Burma

Louisa Charmaine Benson Craig (1941-2010)

Louisa Charmaine Benson Craig from Burma (Myanmar) was a Karen[18] rebel leader of Jewish origin (Cochin and Sephardic Jew). She was the daughter of Saw Benson from an established Cochin, India Jewish family and Naw Chit Khin, a Karen woman. Benson Craig won the title of Miss Burma in both 1956 and 1958. She also represented Burma for the first time in the 1956 Miss Universe contest.

Louisa Charmaine Benson Craig

---

[18]   Karen is a minority population of Sino-Tibetan language that resides in the south of Burma.

CHAPTER 3

Creation of the State of Israel and Evolution of the Israeli–Asian Relationship Since the 1950s

A rich common history has reinforced the deep bonds between Israel and Asian countries. As this chapter illustrates, a unique multifaceted political, military, economic, and cultural relationship has evolved over the years.

Asian Influence on Pre-Independence Israel

The Burma Road in Jerusalem

During World War II, the "Burma Road," which linked Burma to Southwest China, proved a vital corridor for providing supplies to help China resist Japan. In 1948, during Israel's War of Independence, leading figures, including future Prime Minister David Ben-Gurion, Amos Horev, a key operational officer and future President of the Technion Institute of Technology, and General Mickey Marcus, were inspired to build a similar road to break the Arab blockade of Jerusalem.

Jerusalem's equivalent was constructed within a few weeks, passing through hills and mountains in several locations (today Moshav Bekoa, Bait Jiz, Bait Sussin-Kibbutz Harel, Beit Meir) to connect to the old Jerusalem road. After its completion, the besieged Jewish population of Jerusalem was able to receive critical supplies.

Israel and Burma continued to entertain strong ties during the 1950s. The original Burma Road in Jerusalem – or part of it – has been preserved for historical walks and educational purposes.

Construction of the Burma Road in Israel (1948)

Recognition of Israel by Asian Countries

In May 1949, a year after its independence, Israel was admitted as a member of the United Nations, having declared independence one year earlier. Asian countries recognized the State of Israel fairly early and established diplomatic relationships accordingly. The Republic of China[19] recognized Israel in March 1949 and although Israel recognized the People's Republic of China (PRC) in January 1950, official diplomatic relationships were established only in January 1992.

Israel was recognized by the Philippines in May 1949; by India in September 1950; by Thailand in September 1950; by Japan in May 1952; by Burma in July 1953; by Laos in February 1957; by Cambodia in August 1960; by The Republic of Korea in April 1962; by Singapore in May 1969; and by Vietnam in July 1993. Indonesia has not yet officially recognized the State of Israel, although Israelis can travel in the territory with an invitation from the Ministry of Immigration, and large Israeli companies operate discreetly in the country.

---

[19]   The People's Republic of China was established in October 1949.

Front Page of *The Palestine Post*, Israel (1948)

Evolution of Israel and Asian Relationships Since the 1950s

China

In January 1950, Israel was the first country in the Middle East to recognize the People's Republic of China (PRC). During the 1950s, Israel and the PRC nourished political and trade dialogue. High-level meetings were organized both at political and trade levels. In 1954, an Israeli delegation led by Israeli politician David Hacohen and members of the Ministry of Foreign Affairs' Asian department met China's Vice Minister of Foreign Affairs, Zhang Hanfu, and the Director of the Asian & African Department, Huang Hua.

Although the 1960s and 1970s saw a lull in relations, the relationship revived in the 1980s. Israel began to provide China with defense and military equipment. Trade, notably in technology and initially via Hong Kong, started to boom.

In 1985, China admitted Israeli scholars to conferences and symposia. Reciprocally, Chinese scholars started visiting Israel. A direct-dial telephone system was established between the two countries in 1986. Israeli tourists were allowed to travel in China. A delegation of the Communist Party visited Israel in 1987, and Israel's Prime Minister Shimon Peres appointed Amos Yudan to set up the first Hong Kong-based Israeli government-affiliated company to encourage commercial activity between Israel and China. A year later, a delegation of the Israel Labor Party visited China.

The formal establishment of diplomatic relations with Israel in 1992 resulted in a dramatic increase in ties between China and Israel in various political, commercial, and cultural fields. Israeli Defense Minister Moshe Arens visited China in 1991 and negotiated the expansion of military cooperation between the two countries. Israeli Minister of Foreign Affairs, David Levy, visited China in 1992. He appointed Zev Sufott as Israel's first Ambassador to China and Dan Catarivas – previously based in Hong Kong – as Economic Counselor of the Israeli Embassy in Beijing.

In the 2000s, the Israel–China relationship continued to strengthen. The visit of Israeli Prime Minister Ehud Olmert to China in 2007 further boosted trade and military cooperation. In May 2013, Prime Minister Benjamin Netanyahu traveled to China to consolidate and further expand the countries' cooperation and, in return, Chinese Foreign Minister Wang Yi visited Israel that December. Shimon Peres, by then President of the State of Israel, visited China in April 2014 and met with President Xi Jinping.

Over the past twenty years, trade between Israel and China has skyrocketed to an estimated US$9 billion in 2014[20] from US$50 million in 1992. China is Israel's largest trading partner in Asia and the second-largest in the world after the United States. Israel provides China with technology and innovation expertise, specifically in the fields of water, irrigation, desalination, agriculture technologies, renewable energy, cyber security, Internet, telecom, digital media, and healthcare. China provides Israel with numerous manufactured goods.

At an anecdotal level, evidence of the burgeoning relationship is evident. During the authors' first trips to Israel with Chinese investors, it was obvious that even five-star hotels did not cater to

---

[20]   All bilateral figures are provided by Israeli government sources.

Chinese clients. Israeli breakfasts are abundant and luxurious, but definitely have nothing in common with Chinese breakfasts. Hummus, for example, is delicious, but has never been part of a Chinese traditional breakfast. Recently, the same hotels have created new sections in their huge buffets dedicated specifically to Chinese-style food and preferences. Interestingly, one of the descendants of the Kadoorie family living in Israel, Hila Solomon – who specializes in hosting leading politicians and executives in Jerusalem – is now known among Chinese VIPs as one of the best caterers in the city.

Similarly, requests to serve hot water for our Chinese visitors would initially be greeted with incomprehension. Today the majority of hotel restaurants often put a pot of hot water on the meal table for Chinese guests.

India

India recognized the State of Israel in September 1950, but only formally established diplomatic relations with Israel in 1992. Initial relations nurtured in the 1960s were reinforced in the 1970s when Israel provided military support to India during its wars against Pakistan in 1965 for the control of Kashmir, and again in 1971, in the lead up to the independence of Bangladesh. Israel was, in fact, one of the first countries to recognize Bangladesh.

In 1977, Israeli Foreign Minister Moshe Dayan visited the Janata government in India to develop bilateral activity. Prime Minister Rajiv Gandhi met Prime Minister Shimon Peres in 1985 at the United Nations' General Assembly. The Israel–India relationship only really started to take off in the 1990s, however, with India's economic liberalization. In January 1992, India and Israel established official diplomatic relations and in 1997 Ezer Weizman became the first Israeli president to meet Indian President Shankar Dayal and Prime Minister H.D Deve Gowda.

In the following years, the Secretary to the Indian Prime Minister, Soli Sorabjee, and several other Indian officials visited Israel. Throughout the 2000s, the political relationship intensified dramatically with numerous visits of Israeli officials to India. This would include visits from most of the Israeli political establishment (presidents, prime ministers, and ministers of foreign affairs, defense, industry and trade, agriculture, science and technology, education, and transport) as well as political figures such as Shimon Peres, Ariel Sharon, Ehud Olmert, Ehud Barak, Silvan Shalom, Yuval Steinitz, Gabi Ashkenazi, Uzi Landau, Limor Livnat, Eliyahu Yishai, Israel Katz, Shaul Mofaz, Meir Sheetrit and Benjamin Ben-Eliezer. From the Indian side, numerous delegations traveled to Israel over the same period. These included visits from the Ministers of Commerce & Industry and the Minister for Agriculture & Consumer Affairs. In 2012, Foreign Minister S.M Krishna met with Israeli President Shimon Peres, Prime Minister Benjamin Netanyahu, and Foreign Minister Avigdor Lieberman. In September 2014, Netanyahu and Indian Prime Minister Narendra Modi met in New York and discussed the two countries' increasing economic cooperation.

Alongside dynamic political developments, economic cooperation has flourished over the past two decades. Bilateral trade has risen to about US$5 billion in 2014 from US$200 million in 1992. This boom was made possible by the implementation of several trade agreements between India and Israel in the fields of agriculture (1993), air transport (1994), investments (1996), industrial and technology R&D (1996), health and medicine (2003), and environment (2003). Today, India is Israel's second-largest Asian economic partner, and Israel is India's ninth-largest economic partner. Israeli exports to India include diamonds and precious stones, defense equipment, chemicals and mineral products, and high-tech equipment. India meanwhile exports diamonds and precious stones, textiles, metals, plants, and vegetables to Israel.

Cultural exchanges have also been fostered. Israeli and Indian artists have performed together. Famous Israeli singer Achinoam Nini, or Noa, and Israeli Arab singer Mira Awad, have given several concerts in India. Tourism has been highly promoted and in 2014 more than 50,000 Israeli tourists, including 40,000 young Israelis finishing their military service, visited India. A similar number of Indian tourists visit Israel every year and collectively comprise the largest contributor to tourism from Asia.

Singapore

Singapore and Israel had a longstanding relationship even before the official recognition of Israel by Singapore in May 1969. In fact, when Singapore declared its independence in August 1965, its Defense Minister, Dr. Goh Keng Swee, contacted the Israeli Ambassador to Thailand, Mordechai Kidron, to request Israeli assistance in setting up the Singaporean Army. Singapore's "best-kept secret", which was revealed only forty years later by Prime Minister Lee Kuan Yew in order to preserve the confidentiality of the operation, empowered Israeli advisors – who were renamed "Mexicans" for security reasons – to establish the Singapore Army. Until today, Israel and Singapore share a very close relationship in the field of defense.

Over the years, Singaporean and Israeli officials have made numerous reciprocal visits and greater cooperation has been encouraged. Several bilateral agreements have been signed, including, in 1970, in the field of aviation and, in 1997, in the field of research and development with the creation of the Singapore–Israel Industrial Foundation sponsored jointly by Israel's Office of the Chief Scientist and the Economic Development Board of Singapore. Today, Israel exports mainly high-tech equipment to Singapore and Singapore exports machinery and computer equipment to Israel. In 2014, Israel and Singapore's bilateral trade relationship amounted to approximately US$1.5 billion.

Japan

Japan recognized Israel in 1952 and opened an Embassy in Israel in 1963. During the 1960s and 1970s, the relationship between the two countries blossomed. In 1971, the first bilateral agreement regarding visa exemption was signed. Since then, the diplomatic relationship has been very active with the reciprocal visits of prime ministers and ministers of foreign affairs, trade, science and technology, and agriculture. From the 1990s to 2000s, several additional bilateral treaties were signed in the fields of tax law (1993), culture and education (1993), cooperation in science and technology (1995), air services (2000) and research and development (2014). Cultural programs and exchanges have also intensified over the past thirty years, with youth exchange programs (1987), a cultural festival (1992), and Israeli Philharmonic concerts in Japan (2000 and onwards).

In 2014, Israeli Prime Minister Benjamin Netanyahu visited Japanese Prime Minister Shinzo Abe to boost Israeli–Japanese economic cooperation and in 2015 Japan's Prime Minister flew to Israel with a delegation of 100 businessmen to develop trade between the two countries. Bilateral trade between Israel and Japan reached approximately US$1.9 billion in 2014, with Israel exporting US$720 million in machinery, electrical and medical equipment, chemicals, and diamonds and Japan exporting US$1.1 billion in automobiles, chemicals, machinery, and electrical equipment.

Israeli Prime Minister Benjamin Netanyahu
writing a letter on a new Sefer Torah in Tokyo, Japan

South Korea

The Republic of Korea recognized Israel in 1962. Israel opened its embassy in Seoul in 1968. Since the 1960s, Israeli and South Korean officials have engaged in reciprocal visits, beginning with the visit of Israeli Chief of Staff Yitzhak Rabin to South Korea in 1961. Israel notably helped South Korea establish its agriculture, water, and defense infrastructures.

In the 1970s and 1980s, for political reasons, the relationship decreased in intensity and Israel's Ambassador to Japan served as a non-resident Ambassador to South Korea. In the 1990s, however, the Israeli and South Korean relationship dramatically improved with the reopening in 1991 of the Israeli Embassy in South Korea and the opening in 1993 of the South Korean Embassy in Israel. In 1993, South Korea's Science and Technology Minister visited Israel and signed several agreements with the Israeli Ministry of Science, the Weizmann Institute, and the Israel Aircraft Industry.

In 1994, Yitzhak Rabin, then Prime Minister of Israel, visited South Korean President Kim Young Sam. This led to several bilateral agreements, in aviation and culture (1994), science and technology (1994), visa and immigration (1995), and tax (1997). In 1997 and 1999, Rabin and South Korean Prime Minister Kim Jong-pil conducted reciprocal visits.

The relationship between Israel and South Korea strengthened further in the 2000s with the reinforcement of political, economic, and cultural ties. High-profile political visits continued. In 2005 South Korea's Foreign Minister Ban Ki-Moon visited Israel and in 2010 Israeli President Shimon Peres visited Seoul, accompanied by a delegation of senior business executives. In 2001, the State of Israel and South Korea established a joint research and development fund and signed a shipping agreement. From US$148 million in the 1990s, the trade relationship soared to approximately US$2 billion in 2014.

Israel exports mainly manufactured goods, electronics, telecoms, biomedical, defense and security equipment, chemicals, and diamonds and precious stones. South Korea exports automotive, consumer electronics, home appliances, and telecoms products.

Cultural and cooperative exchanges have also blossomed with the rapid development of tourism and the signing of an agreement between Israel's Technion Institute of Technology and Korea Electronics Technology Institute. Christian pilgrims[21] from South Korea have made more than 40,000 visits each year to the Holy Land.

Taiwan

Before being overthrown on the Chinese mainland, the Republic of China recognized Israel in 1949 but due to Israel's relationship with the People's Republic of China, an Israel Economic & Cultural Office (ISECO) was only established in Taiwan in 1993.

From the 1950s to the 1990s, official bilateral activities were limited. In the 1970s, Yaakov Liberman, the President of Taiwan's Jewish community, was also a representative of the Eisenberg Group, which represented Israeli companies in Taiwan. In the 2000s, several bilateral agreements were signed for visa and immigration, customs, and tax (2011). Since then, a trade relationship has developed, reaching almost US$1.2 billion in 2014. Israel's principal exports to Taiwan include chemicals, semiconductors, diamonds, software, electronics, medical, and telecoms equipment. Taiwan exports chemicals, manufactured goods, consumer electronics, and telecoms equipment to Israel.

Burma (Myanmar)

Burma recognized Israel in 1953 and a very early and strong diplomatic relationship was established with Burmese officials. In 1955, Burma's Prime Minister, U Nu, was the first foreign Prime Minister to visit the new Jewish State and in 1961, Israeli Prime Minister David Ben-Gurion visited Burma. In the years that followed, many Israeli officials visited Burma, including Israeli President Yitzhak Ben-Zvi and Foreign Ministers Moshe Sharett, Golda Meir, Abba Eban, Moshe Dayan, and Shimon Peres. From the mid-1960s through the 1980s, the pace of the relationship slowed due to political

---

[21]   One-third of South Koreans are Christians who are familiar with the Bible.

concerns, but political, economic and cultural relationships blossomed once again from the 1990s onward. Israel and Burma developed bilateral cooperation in the fields of agriculture, health, education, commerce, science and technology. The Israel Agency for International Development Cooperation (the MASHAV) has been particularly active in Burma, where it has provided in-depth training to Burmese farmers in crop optimization, including livestock, vegetable, fruit, and flower crops.

Israel and Burma have also cooperated in the field of education through a program developed with UNICEF to support early childhood. In 1994, the countries signed a bilateral agreement on protection of investment, and bilateral trade has since developed rapidly to a level of US$14 million in 2014 from US$2 million in 2012.

Israeli Prime Minister David Ben-Gurion dressed in traditional Burmese clothing during a visit to Rangoon, Burma

Indonesia

Indonesia has not yet recognized Israel diplomatically. The two countries have, however, conducted a series of high-profile meetings. In 1993, Israeli Prime Minister Yitzhak Rabin met Indonesian President Suharto in his Jakarta-based private residence. In 2005, Israeli Foreign Minister Silvan Shalom met with Indonesian Foreign Minister Hassan Wirayuda at a United Nations summit. In 2006, Amos Nadai, the Israeli Deputy Director General for Asia at the Minister of Foreign Affairs and Israeli Ambassador to Thailand, Yael Rubinstein, attended the Economic and Social Commission for Asia and the Pacific (ESCAP) in Jakarta. In 2008, an agreement was signed between Israel's national emergency medical services, Magen David Adom (MDA), and Indonesia's Muhammadiyah Association for the provision of humanitarian emergency health services.

In 2010, an Indonesian–Israel trade bureau was launched to facilitate business cooperation between Indonesian and Israeli businessmen. Although no official bilateral trade figures are available, several private companies have been active in expanding relations. Israeli Koor Trade company has been active in Indonesia, while geothermal company Ormat Industries has provided US$250 million in energy conversion technology for the landmark US$1 billion, 330-megawatt Sarulla electricity project.

The Philippines

Israel was recognized by the Philippines in 1949. Even before that, in 1947, the Philippines was one of the thirty-three nations – and the only Asian country – that supported the creation of the State of Israel at the United Nations. In 1958, full diplomatic relations were established and, in 1962, embassies were opened in both countries. From the 1960s through the 2000s numerous high-profile official delegations visited both Israel and the Philippines. In 2012, Philippines Vice President Jejomar Binay visited Israeli President Shimon Peres, and in 2014 Philippines Defense Secretary Voltaire Gazmin also came to Israel.

Several bilateral agreements have been signed between the two countries in air services (1951), agriculture cooperation (1964), visa and immigration (1969), education and academic research (1989), science and technology (1992), tax law (1993), social security (2009), and aviation agreements (2013). Bilateral trade is estimated at US$290 million in 2014, with Israel exporting approximately US$250 million in electronics, chemicals, machinery, consumer products, textiles, and construction materials, and the Philippines exporting approximately US$40 million of electronics, chemicals, marine products, and processed food.

Both Israeli defense companies and the MASHAV are active in the Philippines and offer diverse courses in modern agriculture and clean technologies. The cultural relationship between Israel and the Philippines is also flourishing, with Israel Film Festivals in Manila, visiting Israelis and Filipino artists performing in various music festivals in the Philippines and Israel, and a handicraft and food week festival in Haifa, Israel. An estimated 30,000 Filipinos are living in Israel, mostly as foreign workers, and numerous Christian pilgrims visit the Holy Land each year.

Thailand

Thailand recognized Israel in 1950 with full diplomatic relations being established in 1954. An Israeli Embassy was inaugurated in Bangkok in 1958 and a Thai Embassy opened in Tel Aviv in 1996. Over the years, high-profile Thai officials have visited Israel, such as Crown Prince Vajiralongkorn and Princess Maha Chakri Sirindhorn; likewise, Israeli dignitaries, including Chief of Staff Yitzhak Rabin and Abba Eban, have visited Thailand. Several bilateral agreements have been signed in various fields, including air services (1968), tax law (1996), protection of investment (2000), and irrigation technology (2002). The MASHAV is active in Thailand and offers training in agriculture for Thai professionals. In 2014, the KU[22]–Israel Agriculture Technologies

---

[22] Kasetsart University.

Complex opened using Israeli agro-technology solutions to optimize Thai crop production.

Trade between Israel and Thailand is estimated at approximately US$1 billion in 2014, with Israel exporting US$500 million to Thailand in diamonds, machinery, fertilizers, and chemicals, and Thailand exporting US$500 million in diamonds, machinery, vehicles, and cereals. Strong Israeli and Thai cultural relationships have also developed in music, crafts and food, and more than 100,000 Israeli tourists visit Thailand each year[23]. Israel hosts a large Thai population, especially agricultural workers.

Vietnam

Vietnam recognized Israel in 1993. Israel subsequently opened an embassy in Hanoi, and in 2009 Vietnam inaugurated its embassy in Tel Aviv. A series of high-profile visits has nourished the diplomatic relationship. In 2011, Vietnamese Minister of Defense Truong Quang Khanh was received in Israel and the visit was reciprocated by Israeli President Shimon Peres.

Israel and Vietnam cooperate in the fields of agriculture, water, telecoms, information services, and homeland security. Israel and Vietnam have signed a series of bilateral treaties in the fields of economic cooperation (2004) and tax law (2009). The MASHAV is active in Vietnam and offers training in agriculture for Vietnamese professionals. In 2014, bilateral trade between Israel and Vietnam amounted to some US$630 million. Cultural relationships have developed in various fields; in 2014, for example, an Israeli Film Festival was inaugurated at the National Cinema Center in Hanoi.

Cambodia and Laos

In 1955, Israel recognized both Cambodia and Laos. Cambodia recognized Israel in 1960 and established diplomatic relations in 1993. Laos recognized Israel in 1957 and also established a diplomatic relationship with the Jewish state in 1993. Since the 1950s, high-level

---

[23]   Figures provided by the Thailand Ministry of Tourism.

officials in Israel, Cambodia, and Laos have exchanged visits. In 1956, Israel's Foreign Minister, Moshe Sharett, visited Cambodia and in 1962, Israeli Foreign Minister Golda Meir visited Cambodia's King Norodom Sihanouk. In 1974, Yitzhak Rabin visited Laos. Over the years diplomatic relationships between Israel, Cambodia, and Laos have intensified dramatically. In the economic arena, Israel has been active in Cambodia and Laos in the fields of agriculture, high-tech, homeland security, water, and irrigation. Cambodia exports commodities such as rice and tobacco leaves, while Laos' exports to Israel are still in their infancy. Economic and cultural relationships between Israel, Cambodia, and Laos have improved in recent years and are expected to develop further.

PART TWO

ISRAEL WORLD INNOVATION CENTER

CHAPTER 4

Israel's DNA: Science, Technology, and Innovation

Nurturing Innovation

Research, science, technology, and innovation have characterized the modern history of the Jewish people, particularly the early settlers who came to the Holy Land before the establishment of the State of Israel. With scarce resources and a challenging climate and environment, scientific innovation has been used as a tool to enhance the lives of Israel's citizens.

The first Israeli research institutions were created in the fields of agronomy and agriculture. Mikveh Israel School was inaugurated in 1870 by the Alliance Israelite Universelle, and Agriculture Research Organization was founded in 1921 in Rehovot. The same pioneer spirit has nurtured Israeli research and innovation over the decades. With its leading universities, research institutions, and elite military units, Israel has become an internationally recognized powerhouse in technology and innovation.

Leading Israeli Universities and Research Institutions

Technion Institute of Technology

The Technion Institute of Technology is Israel's oldest university. It was created in 1912, with the help of the German Ezra Fund, as many Jews were not allowed to access universities and science courses in Europe. In 1923, Albert Einstein, the distinguished physicist who developed the theory of relativity, visited the Technion in Haifa and served as the first President of the Technion Society. The following year, the Technion opened its doors to students.

The Technion's numerous faculties include aerospace engineering, architecture and town planning, biology, biomedical engineering, biotechnology and food engineering, civil and environmental engineering, chemical engineering, chemistry, computer science, education in technology and science, electrical engineering, humanities and arts, industrial engineering and management, materials sciences and engineering, mathematics, mechanical engineering, medicine, and physics. The Institute has over 14,000 students with more than 9,500 undergraduates, 2,400 postgraduates, 1,000 doctoral students, and 52 research centers.

The Technion established a technology transfer bureau, Technion Technology Transfer (T3), in 2007 to commercialize in-house research to investors. In 2010, the Technion won a bid to establish a science and engineering school at Cornell University in the United States. In 2013, with the help of Li Ka Shing Foundation, it established an institute of technology with Shantou University in China.

The Technion Institute of Technology regularly ranks among the top fifty universities in the world, especially for chemistry, computer science, mathematics, natural sciences, and engineering. Four Technion faculty members have received the Nobel Prize in Chemistry: Avraham Hershko and Aaron Ciechanover (2004), Dan Shechtman (2011), and Arie Warshel (2013). Technion alumni include numerous entrepreneurs and CEOs of Israeli and international firms, such as Dadi Perlmutter (Chief Product Officer of Intel), Abraham Lempel and Jacob Ziv (developers of the Lempel-Ziv, LWZ, compression algorithm), Andi Gutmans (developer of PHP and co-founder of Zend Technologies), Yehuda and Zohar Zisapel (co-founders of the Rad Group and patrons of the Israeli high-tech industry) and Yoelle Maarek (a founder of Google Israel and CEO of Yahoo Labs Israel). Both the Iron Dome missile defense system and the ReWalk robotic exoskeleton that enables paraplegics to walk are innovations of Technion alumni.

Hebrew University

The Hebrew University was founded in 1918 and opened in 1925. The university's first Board of Governors included Albert Einstein; Sigmund Freud, known as the father of psychoanalysis; the philosopher Martin Buber; and Chaim Weizmann, who later became the first president of Israel. The University has six campuses, seven faculties (humanities, social sciences, law, mathematics and sciences, agriculture and food environment, medicine, dental medicine) and fourteen schools. It has approximately 23,000 students – of which over 45% are postgraduates – and some 83 research centers.

The Hebrew University is among the top 50 universities in the world. It has a particularly strong reputation in the fields of mathematics, computer science, and business and economics. In 1964, the university created its technology transfer bureau, Yissum Research Development Company, which is in charge of the commercialization of Hebrew University's intellectual property to private investors. Yissum enabled nine companies to form based on its in-house technology in 2014. In the same year, sixty-five technology licensing agreements and about 600 technology-based assignments were concluded.

Nearly 40% of all civilian scientific research in Israel is conducted at the Hebrew University. Seven Hebrew University researchers and alumni have received the Nobel Price, including Daniel Kahneman in economics (2002), David Gross in physics (2004), Avram Hershko in chemistry (2004), Aaron Ciechanover in chemistry (2004), Robert Aumann in economics (2005), Roger Kornberg in chemistry (2006) and Ada Yonath in chemistry (2009). Among the university alumni, three were Israeli Presidents: Ephraim Katzir, Yitzhak Navon and Moshe Katsav. Three more – Ehud Barak, Ariel Sharon and Ehud Olmert – were Israeli Prime Ministers. Numerous other alumni became leading CEOs, including Léo Apotheker (former CEO of HP and SAP), Orit Gadiesh (former Chairman of Bain & Co), and Eli Hurvitz (former CEO of Teva Pharmaceuticals).

Weizmann Institute of Science

Chaim Weizmann, who subsequently became Israel's first President, founded the Daniel Sieff Research Institute in Rehovot in 1934. It was renamed The Weizmann Institute in 1949.

Weizmann was born in Belarus. He acquired British citizenship in 1910 and held it until 1948 when he became President of the new state. An accomplished biochemist, he developed the technique of creating acetone through bacterial fermentation. This was a major contribution to Britain in World War I and contributed to Weizmann's great stature and influence in Great Britain.

In 1955, Weizmann Institute of Science developed WEISAC, the first modern computer in the Middle East and one of the world's first electronic computers. In 1959, the Institute created a technology transfer bureau, Yeda Research and Development Company, to promote its intellectual property to industrial applications. Among its successes are Bio-Hep B®, a recombinant hepatitis B vaccine, the world's first commercial solid lubricant based on spherical inorganic nanoparticles, and improved crop varieties and hybrid cultivars of cucumbers and melons.

Today, the Weizmann Institute of Science offers only graduate and post-graduate courses. It has five faculties involved in the fields of mathematics and computer sciences, physics, chemistry, biochemistry, and biology. The institute has more than 1,100 students, 700 doctoral and 400 post-graduate students, and more than 50 research centers. Weizmann Institute of Science has three Nobel Prize winning professors: Ada Yonath in chemistry (2009), and Michael Levitt and Arieh Warshel in chemical physics (2013). The institute ranks among the top fifty academic institutions worldwide with particular strengths in the fields of computer science and chemistry.

Tel Aviv University

Tel Aviv University was created in 1956 when the Tel Aviv School of Law and Economics, and the Institutes of Jewish Studies and Natural Sciences merged. Tel Aviv University now has nine faculties (arts, engineering, exact sciences, humanities, law, life science, medicine, social sciences, management) and seven schools (environmental studies, music, architecture, dental medicine, education, social work, international school). Tel Aviv University has more than 31,000 students, 3,300 doctoral students, and 130 research institutes. It regularly ranks among the top fifty academic institutions in the world, particularly in the field of computer science. Alumni include the former Prime Minister of Israel Ariel Sharon, Israel's first astronaut Ilan Ramon, and Waze co-founders Uri Levine and Ehud Shabtai.

Ben-Gurion University of the Negev

David Ben-Gurion, Israel's first prime minister, believed that the South would play a critical role in the destiny of the state, since this area, which is largely desert, covers 60% of Israeli territory. Ben-Gurion University of the Negev was created in 1969 to help fulfill that vision. Today, the university has more than 20,000 students, of which 13,000 are undergraduates, 1,300 doctoral students, and 6,700 post-graduates. Ben-Gurion University has three campuses in Beer Sheva and two in Sde Boker and Eilat. It has five faculties in sciences (health sciences, natural sciences, business management and humanities and social sciences), and six schools (graduate studies, medical school, community health, pharmacy, medical laboratory sciences, continuing medical education). Eight research institutes and more than sixty-one interdisciplinary research centers, including the Solar Energy Center and Institute for Water Research, highlight the university's strong emphasis on research.

BGN Technologies, Ben-Gurion University's technology transfer bureau, monetizes the university's intellectual property. Ben-Gurion University ranks as one of the top in the world in the fields of water and solar research. Alumni include Former Minister of Finance, Science and Foreign Affairs Silvan Shalom and former CEO of Israel airline El Al Eliezer Shkedi.

Bar-Ilan University

Bar-Ilan University was created in 1955. Today it has more than 26,000 students, of which approximately 17,000 are undergraduates,

2,000 doctoral students, and 6,800 post-graduates. Bar-Ilan University has nine faculties and more than ten research centers. It also houses Israel's largest nanotechnology center.

The university's technology transfer bureau, Bar-Ilan Research and Development Company (BIRAD), monetizes Bar Ilan's intellectual property to private investors. Bar-Ilan University, and especially its computer sciences program, ranks among the top 100 institutions in the world. University alumni include former Minister of Justice Tzipi Livni and Israeli politician Michael Ben-Ari.

Leading Israeli Research Centers

Volcani Institute of Agricultural Research

Yitzhak Volcani founded the Institute of Agricultural Research in 1921 in the city of Ben Shemen as an agricultural experiment station. In 1932, the station was relocated to Rehovot in the Tel Aviv area. When Volcani passed away in 1951, the Israeli Minister of Agriculture assumed the management of the center, renaming it in honor of its founder.

The Volcani Institute of Agricultural Research later moved to Beit Dagan, also in the Tel Aviv area, and became part of the Agricultural Research Organization (ARO) under the umbrella of the Ministry of Agriculture and Rural Development. ARO has six institutes and three campuses under management. The Volcani Institute specializes in agriculture under arid conditions, irrigation using recycled wastewater and desalination, freshwater aquaculture under water shortages, enhancement of crop optimization and production. The center is active in six areas, including plant protection and sciences, animal sciences, soil, water and environmental sciences, agricultural engineering, and food sciences. The Institute of Agricultural Research provides training to graduate students in cooperation with Israel's leading universities. It also aids agricultural workers in developing countries under the leadership of the MASHAV. In addition, through its technology transfer bureau Kidum, ARO monetizes its intellectual property to third-party investors.

Israel Institute for Biological Research

Prime Minister Ben-Gurion's science advisor and Head of Research and Development at the Ministry of Defense, Professor Ernest Bergmann, created the Israel Institute for Biological Research in 1952. The institute, which operates under the jurisdiction of the Prime Minister's office, specializes in vaccines and pharmaceuticals, protein and enzymes, medical diagnostics, biotechnology, and environmental detectors. Its technology transfer bureau, Life Science Research Israel, commercializes the research center's intellectual property.

Israel Defense Forces' Specialized Units and Programs

The Israel Defense Forces' technology units form a critical component of Israel's innovation capability. As a country facing constant threats from its neighbors, defense is of even greater importance than in many other countries. Israel has put a high priority on innovation in defense technology to maintain a competitive edge.

Mamran Unit

Mamran, an acronym for Center of Computing and Information Systems, is the Israel Defense Forces' (IDF) central unit dealing with computing systems. The unit was formed in 1959, hosting several data processing units, such as the inventory processing center and manpower computer center. Today, Mamran provides data processing services for the IDF. Its responsibilities include management of the IDF's internal intranet, network systems, and integration.

Considered one of the best computer programming institutes in the world, Mamran created a dedicated school called Basmach in 1994. In 2007, the Mamran Association was launched to provide its members with networking opportunities. Mamran alumni are the human resources backbone of Israel's high-tech industry. Many have become influential CEOs of large Israeli corporations, such as the former CEO of Bank Leumi, Galia Maor, and the CEO of Matrix IT, Moti Gutman.

8200 Unit

The 8200 Unit, created in 1952, is responsible for collecting signal intelligence. It is the largest unit in the IDF and is comparable to the United States' National Security Agency, collecting military intelligence from diverse sources. Since its inception, this unit has been providing the manpower that enabled the creation of the Israeli high-tech industry. 8200 alumni are founders and CEOs of numerous Israeli technology companies, including Check Point, ICQ, Nice Systems, Gilat, Audiocodes, and EZ Chip.

Talpiot Program

The Talpiot Program, created in 1979, enables its recruits to pursue a university degree in the sciences while serving the IDF in a research and development position. Alumni of the program include founders of Compugen, Eli Mintz, Simchon Faigler, and Amir Natan; one of the co-founders of Check Point Software, Marius Nacht; and Chief Technology Officer of Trusteer, Amit Klein.

9900 Unit

The 9900 Unit, almost unknown to the general public, is a special intelligence unit that gathers valuable visual information from maps to satellites images. The unit's members analyze route movement and field data. The unit develops its advanced technology, such as sensors and augmented reality devices, internally. VisionMap's digital 3D photography civilian application, one of the unit's success stories, was acquired by Israel's Rafael and South Korea's SK Group in 2013 for US$150 million.

Israel's Innovation and International Recognition

Recognition by International Institutions

Nobel Prize

Twelve Israelis have won the Nobel Prize, the majority in the field of chemistry. These include Avram Hershko and Aaron Ciechanover for their discovery of ubiquitin-mediated protein degradation (2004), Ada Yonath for her studies of the structure and function of the ribosome and how cells build proteins (2009), Dan Shechtman for the discovery of quasi-crystals (2011), and Michael Levitt and Arieh Warshel for the development of a multi-scale model for complex chemical systems (2013). The regularity with which Israelis win the Nobel Prize in chemistry is recognition by the international scientific community of Israel's leadership in this branch of scientific endeavor.

Nobel Prizes have been awarded in the field of economics to Daniel Kahneman (2002) and Robert Aumann (2005); and to Shmuel Yosef Agnon in Literature (1966). Menachem Begin (1978), Shimon Peres (1994), and Yitzhak Rabin (1994) have meanwhile all won the Nobel Peace Prize.

Fields Medal

The Fields Medal, known as the International Medal for Outstanding Discoveries in Mathematics, was awarded to Elon Linderstrauss in 2010 for his results on measure rigidity and their application to number theory.

European Research and Development Programs

Israel has participated in various European Framework Programs for Research and Development on an ongoing basis, including, among other things, Galileo global satellite navigation, and the Eureka Network for Market-Oriented R&D.

In 2010, Israel was elected Chairman of the Eureka Program, with the purpose of leading industrial research and development in Europe, even though Israel was the only non-European country taking part. In June 2014, European Commissioner Manuel Barroso signed an agreement with Israeli Prime Minister Benjamin Netanyahu allowing Israel to join the new European Union Research and Innovation Program, Horizon 2020. The agreement grants Israel the same access as EU Member States and associated countries with the same budget contribution obligation.

European Organization for Nuclear Research — CERN

CERN, which operates the largest particle physics laboratory in the world, accepted Israel as an associate member in 2011. Israel is the first non-European country to gain admission. More than forty Israeli scientists work within the organization.

International Recognition for Landmark Achievements

Israel's contribution to science, technology and innovation has led to global recognition in many fields. It registers the largest number of high-tech start-ups after the United States, the highest number of engineers per capita, and has been recognized as world's second most highly educated country[24]. The OECD ranks Israel first in civilian research and development spending as a percentage of GDP with 2.26%[25] – double the average of OECD countries. In 2013, Israel invested a massive 4.38% of its GDP in research and development, the highest percentage in the world[26] and is ranked second in the Global Dynamism Index for science and technology[27]. In 2014, Israel ranked first for innovation capacity[28], first for entrepreneurship and third for global innovation. In 2014, Israel also ranked first in the Global Cleantech Innovation Index[29] and fifth for world patent filings per capita[30]. The

---

[24]   OECD Education report (2012).
[25]   OECD Science & Innovation report (2008).
[26]   OECD Science and Technology Report (2013).
[27]   Grant Thornton Index (2013).
[28]   IMD Global Competitiveness Yearbook (2014).
[29]   Cleantech Group and World Wildlife Foundation (2014).
[30]   World Economic Forum Global Competitiveness Yearbook (2014-2015).

quality of Israeli research institutions as a whole was ranked third in the world.

Landmark Achievements

Israel has been at the forefront of global technology. With its strong research and development capacity, leading universities and research centers, and innovative army, Israel has not only enabled the emergence of internationally recognized researchers, but has also empowered various local industries to achieve breakthrough innovations.

To name just a few, these include microprocessors with revolutionary consumption and cost efficiency, such as Intel's Centrino and Core 2 Duo products, USB Flash drives by M-Systems, creation of the first endoscopic capsule camera by Given Imaging, CT scanning by Elsint, IP telephony with Vocaltec, instant messaging with ICQ, and commercial digital offset color printing by Indigo. Recent innovations include Viber's VoIP telephony, Waze's social GPS, and StoreDot's fast-charging batteries.

StoreDot CEO Doron Meyersdorf

In the field of agriculture, in 1973, two Israeli agricultural scientists developed a variety of cherry tomato that ripens more slowly in a hot climate than ordinary tomatoes. Israel has become known for its advanced pomegranate production, with locally developed varieties reaching an average yield of 25-35 tons per hectare with improved quality and taste. Innovative greenhouse systems enable Israeli farmers to grow more than 3 million roses per hectare per season and an average of 300 tons of tomatoes per hectare per season – four times the yield of open fields. Forty percent of European tomato greenhouses use long-life hybrid seeds created in Israel[31].

Israel is also the second-largest global producer of loquats after Japan and one of the world's leading fresh citrus producers. Exports include oranges, grapefruits, and tangerines.

In the cotton industry, Israel has enabled both the shortening of the growing season and larger yields. It has been using greenhouses to optimize horticultural productivity, creating a system that is three times as efficient as traditional methods. As a result, 80% of Israel's flower production is exported to Europe.

In dairy production, innovations include advanced computerized milking and feeding systems, cooling systems to reduce heat stress on cows, embryos for transplant, milk processing equipment, consultancy services, and international project development. Advances in animal husbandry include the development of Assaf sheep with higher milk yields, a result of the crossbreeding of the improved Israeli Awassi and the German East Friesian breeds, as well as prize-winning Holstein cows with superior milk production capabilities. In fact, with an average of 10,000 liters, Israel's cows produce the largest annual volume of milk per animal in the world.

Given that 60% of the country is natural desert, it is hardly surprising that Israel leads the world in water recycling with an 80% recycling rate[32]. It has helped achieve 70%-80% water efficiency in

---

[31]   Israel Ministry of Industry Trade & Labor *Invest in Israel* report (2011).
[32]   The second-largest water recycler in the world is Spain with a rate of 18%.

agriculture, again the highest rate in the world. As a result, Israel has also achieved the highest global ratio of crop yield per water unit.

A world leader in irrigation, Israel offers innovative technologies and accessories. Drip irrigation, automatic valves and controllers, automatic filtration, low discharge sprayers, use of low saline content water, mini-sprinklers, compensated drippers, and sprinklers have enabled numerous industry breakthroughs.

The country is home to the world's largest seawater reverse osmosis desalination plants, annually producing 140-150 million m^3 at a low cost of approximately USD 0.52/m^3, the most cost-efficient of its kind in the world. This has helped Israel overcome a severe shortage of water to become self-sufficient and provide water to a growing population and economy[33]. Israel desalinates 50% of its municipal water. Meanwhile, geothermal water captured underground in the Arava Desert is used to heat greenhouses and fish farms.

In the energy sector, flat solar systems for water heating were optimized in Israel in the 1950s. Israel is the world leader in the use of solar energy per capita with 85% of households using thermal systems. It is also a world leader in new solar technology, such as solar thermal advances developed by BrightSource's Luz.

In several aspects of life sciences, Israel leads the pack. It has the highest number of medical device patents per capita and is fourth in the world for biopharmaceutical patents. Israel is second in Europe per capita for the product pipelines of its biotech companies and first for clinical trials of cell therapy treatments.

In the increasingly important field of homeland security, Israel was responsible for the first unmanned aerial vehicle (UAV) created by Israel Aerospace Industries, the first unmanned naval patrol vehicle developed by Rafael and forward-looking infrared technology invented by Elbit Systems.

---

[33] At one time it was thought that Israel would have to import water from abroad to meet its needs.

International Financial Market Recognition

Other than domestic American firms, Israel accounts for the largest number of companies listed on the NASDAQ, and the third highest number of foreign companies listed in the United States as a whole after Canada and the United Kingdom. Approximately seventy Israeli companies trade on European stock exchanges. Increased international market recognition also results from the coordinated actions of the Israel Advanced Technologies Industries (IATI) organization under the leadership of Karin Mayer-Rubinstein. IATI represents the diverse aspects of Israel's high-tech and biotech industries with a diversified member base ranging from start-ups to multinationals and venture capital firms.

CHAPTER 5

Israel's High-Tech Industry and Sectors of Excellence

Creation and Evolution of Israel's High-Tech Industry

Israel's vibrancy in science, technology, and innovation has allowed it to transform its strength in research and development into a significant component of the economy. Israel has created thousands of high-tech companies, which, combined with medium-tech firms, currently contribute to more than 80% of the country's total exports. How this came about is itself a fascinating story.

Pioneer Years – 1960s-2000s

Israel's High-Tech Backbone

Israel's high-tech industry came to life at the beginning of the 1960s with the incorporation of the Electronics Corporation of Israel in 1961, and Elron and Tadiran in 1962. These three companies formed the backbone of an industry that has since given rise to several thousand new, innovative firms. The technology developed by these original high-tech companies, allied with talented human capital, served as the trigger for Israel's high-tech success story.

Uzia Galil, a Romanian Jew who fled the Nazis and eventually studied engineering at the Technion, created Elron. The company started by producing measurement instruments for medical and electronic applications. In 1966, Elron established Elbit Computer, a joint venture with the Israeli Ministry of Defense, to build minicomputers for defense applications. In 1967, the Elbit 100 computer was launched, and over time Elbit became one of Israel's largest corporations. It spun off in 1996 to create Elbit Systems, a leading defense electronic products company, and Elbit Medical Imaging, which controlled Elsint, a leading medical imaging company.

Elsint was founded in 1969 by Dr. Avraham Suhami, who immigrated to Israel from Turkey at the age of 14 and became a professor of nuclear physics at the Technion Institute of Technology. Financed by Elron, Elsint developed into one of the world's leading medical imaging companies. In 1972, it was the first Israeli company to list on the NASDAQ. Elsint's innovation contributed to a global advancement in medical imaging, including the creation of the world's first multiple CT scanner. In 1999 and 2000, Elsint sold its imaging businesses to GE Healthcare and Philips Medical Systems.

Elron participated in the establishment and development of several Israeli high-tech corporate leaders with global reach. Orbotech, established in 1981, is a leading provider of automated optical inspection and computer-aided manufacturing systems. Zoran, created in 1983, provides complete original equipment manufacturing (OEM) solutions for consumer electronics products. Given Imaging, launched in 1998, develops video capsules with miniaturized cameras for internal non-invasive medical diagnostics.

Tadiran, which acted as an Israeli holding company, set up a number of the country's leading companies.[34]

In addition to Elsint, one of the first Israeli companies to list on the NASDAQ in 1982 was Electronics Corporation of Israel, now ECI Telecom, which specializes in the manufacturing of advanced electronic equipment. ECI Telecom has achieved numerous technological breakthroughs over the years. These include the telephone line doubler (1977); HDSL, one of the earliest versions of DSL (1993); optical ring-based networks (2000); toll-quality voice over IP (2002); coarse wavelength division multiplexing (2003); a carrier Ethernet switch/router built with a transport mindset;

---

[34] Tadiran group companies include Tadiran Telecom; Tadiran Batteries, Tadiran Appliances, Tadiran Telecommunication, Tadiran Systems, Tadiran Spectralink, Tadiran Communication, and Ituran & Tadiran Telematics.

dynamic spectrum management level 3 (2009); and a network design platform (2009).

In the 1970s and 1980s, the RAD Group along with several other Israeli firms, contributed significantly to the rapid development of Israel's high-tech industry. Founded in 1981 by Yehuda and Zohar Zisapel, children of immigrant parents from Poland, the RAD Group formed a unique cluster of independent companies active in the field of telecommunications and networking.[35] Four of these – Radcom, Ceragon Networks, Radware, and Silicom – are listed on the NASDAQ. The RAD Group has empowered more than fifty executives to become serial entrepreneurs and has helped establish hundreds of leading Israeli high-tech firms.

External Factors Influencing the Rise of Israel's High-Tech Profile

Facing intense hostility from most of its neighbors, Israel was forced to develop its military industry even before the establishment of the state. The initial focus was on small arms, including the world famous Uzi submachine gun, which was designed in the late 1940s. Development of the Israeli defense industry, however, accelerated greatly after 1967, following an arms embargo imposed by France in the wake of the 1967 Six-Day War.

As Israel's defense industry flourished in the 1970s, civilian applications for the technology developed became more evident. Scitex Corporation, founded in 1968 by Efraim Arazi who studied engineering at MIT and electronics in the Israel Defense Forces, provides an early success story of civilian adaptation. Its revolutionary digital printing systems employ an adaptation of the fast-rotation drums first used for electronic warfare systems.

In 1985, Scitex established Context Graphics Systems, a joint venture with American firm Continental Can that invented two- and three-dimensional design systems based on Silicon Graphic

---

[35]　RAD group companies include Bynet, Rad Data Communication, Silicom Connectivity Solutions, Radcom, Ceragon Networks, Radware, Radwin, Packetlight Network, RADiFlow, and SecurityDAM.

workstations. In the 1990s, the company was dismantled: Hewlett-Packard purchased Scitex Vision, Eastman Kodak bought Scitex Digital Printing, and Creo Products purchased Scitex Graphic Art Group.

In the 1980s, the global computing industry's shift from hardware to software further accelerated Israel's transition into a high-tech powerhouse. Israel began to develop niche software areas that were not dominated by the United States. From the mid-1980s to the beginning of the 1990s, Israel's annual software exports rose exponentially from US$5 million to US$110 million.

Many of the Israeli software companies formed during this initial boom period remain global market leaders today. The list of well-known corporate names is impressive. Founded in 1982 as Aurec Information, Amdocs is now a leading global provider of software and services to communication, media, and entertainment firms. Magic Software Enterprises, established in 1983, provides software platforms for enterprise mobility, cloud, and application and business integration. Comverse, which began as Efrat Future Technologies in 1983, is a world leader in developing and marketing telecommunication software. Aladin Knowledge Systems, created in 1985, produced software for digital rights management and Internet security prior to its acquisition by American company SafeNet in 2009. Nice Systems, formed in 1986, is a global leader specialized in telephone voice recording, data security, and surveillance. Mercury Interactive, established in 1989 and bought out by HP in 2006, specialized in software for application management, application delivery, change and configuration management, quality assurance, and IT governance. Meanwhile, one of the first firewalls was designed by Check Point Software Technologies, created in 1993, and a leading provider of software and hardware products for IT security.

Pioneering American Multinationals with Israeli R&D Centers

Pioneering multinational corporations, particularly American firms, were quick to understand the advantages of activity in Israel, especially in research and development (R&D). IBM was the first foreign technology firm to establish a presence in Israel, in 1950. The company assembled and repaired punch card machines, opened a plant in 1956, and later offered computerized data processing services. IBM opened its first Israeli R&D center in 1972 in Haifa and still conducts research in the fields of computer science, electrical engineering, mathematical sciences, and industrial engineering. Intel set up its R&D team in 1974, also in Haifa. Among its many accomplishments since then are the development of the first PC processor in 1997, the Pentium MMX, and the release of the Centrino processor.

Spurred on by these early successes, many more firms went on to establish R&D centers in Israel. Google set up an R&D office in 2007 under the leadership of Dr. Yoelle Maarek, a Technion computer science graduate whose team pioneered the "Google suggest" function. Others include HP, Marvell, Yahoo (US), Siemens, SAP (Europe), Samsung, LG, Haier, and Huawei (Asia).

Dot-com Boom and Crisis

The success of a company called Mirabilis was a catalyst for Israeli entrepreneurs, who created thousands of start-ups and successfully raised capital between 1997 and 2000. Mirabilis developed the ICQ instant messaging program that revolutionized communication over the Internet. AOL bought the company in 1996 for US$407 million, only eighteen months after its founding, and even though the company had yet to generate any revenue.

Venture capital raised by Israeli start-ups increased dramatically in 2000 to US$3.7 billion from US$1.9 billion the previous year. Over that period, more than fifty Israeli firms completed an initial public offering on the NASDAQ or other international markets.

Several factors influenced this rise: stronger international media coverage focused on the country's ability to innovate; the arrival of highly skilled immigrants from the former Soviet Union; and the positive prospects for an improved political climate emanating from the Oslo Accords, the ultimately unsuccessful peace agreement between Israel and the Palestine Liberation Organization that was signed in 1993.

The dot-com boom reached its peak in 2000, with the NASDAQ hitting historical highs, before the bubble abruptly popped, dragging down the global technology industry, with blue-chip icon stocks such as Cisco dropping 86%. The dot-com crisis severely affected the Israeli high-tech industry, including the creation of local start-ups. The situation improved dramatically in the years that followed, however, enabling a solid recovery.

The Start-up Nation (2000s-present)

The term Start-up Nation[36] is now broadly used to describe Israel's global influence and leadership in the high-tech industry. Israel's start-up creation in its own Silicon Wadi (*wadi* means valley in Arabic) is second only to the United States' Silicon Valley. Approximately 10,000 high-tech companies have been created in Israel since the 2000s, and 5,400 firms are believed to be active. An increasing number of Israeli companies have listed on American stock exchanges, including seven firms on the New York Stock Exchange, six on the American Stock Exchange and the majority on the NASDAQ, which currently hosts more than seventy-four Israeli corporations. Since 2006, the pace of start-up creation, venture capital, private equity funding, and exits either through mergers and acquisitions or initial public offerings has increased dramatically.

Growing Venture Capital and Private Equity Community

The momentum created by the Start-Up Nation ethos has been given a strong boost by the increasing presence of venture capital and private equity investment funds, seeded by Israel's government-owned Yozma Program and international funds. Injections have come from the corporate venture capital of foreign multinationals (Intel, Microsoft), the increased activity of foreign funds (Index Ventures, Horizon Ventures, Israel Asian Fund), and the development of Israeli funds (Pitango, JVP, Gemini). By 2014, the level of fundraising by Israeli high-tech firms had more than doubled to reach US$3.4 billion

---

[36]   Popularized in the book *Start-Up Nation*, 2009, by Dan Senor and Saul Singer.

raised by 668 companies, compared with US$1.6 billion from 402 companies in 2006.

Fundraising has focused mainly on the Internet (28%), life sciences (24%), and software (22%) sectors. During the global financial crisis, US$1.2 billion was raised in each of 2009 and 2010.

Increasing Number of Initial Public Offerings

The quantity of initial public offerings (IPOs) of Israeli companies in the United States and especially on the NASDAQ, is a clear indicator of the dynamism of Israel's high-tech industry. Since the 1980s, more than 250 Israeli firms have completed IPOs on the NASDAQ. Many of these have since been acquired, merged, or delisted. In 2000, one of the most active years for offerings, more than thirty-three firms listed on the exchange.

Since then, the flow of IPOs has varied, but is still growing. In 2006 alone, for example, major IPOs included broadband optimization firm Allot Communication, software company Perion Network, and security systems firm Supercom. In 2007, they included defense company Acro, telecoms company B Communications, Ethernet and InfiniBand provider Mellanox Technologies, biotech firm OrganiTech, biotech company Rosetta Genomics, satellite service company RRS at Global Communication Network, VoIP solution Veraz Networks, and oil exploration company Zion Oil & Gas.

In 2008 and 2009, IPOs included biotech companies Cell Kinetics and Prolor Biotech, and mobile phone operator Cellcom Israel. The next two years saw the listing of water fountain carbonated systems firm SodaStream, biopharmaceutical company BioLineRX, data security firm Imperva, and quartz surface company Caesarstone. In 2013, they included pharmaceutical companies Alcobra, Kamada, and Oramed Pharmaceuticals; biotech firm Enzymotec; medical equipment firm Mazor Robotics; and Internet page builder Wix.

Israel experienced the highest level of activity for a decade in 2014, both in terms of the number of IPOs (eighteen mostly in New York and London) and the amount raised of US$9.8 billion. This compared with US$1.2 billion raised in 2013. High-profile offerings on the NYSE included Mobileye's advanced driver assistance systems, which raised US$1 billion, making this Israel's largest IPO of the year. NASDAQ offerings included software company CyberArk, and data and access control company Varonis Systems. In London, offerings included payment technology firm SafeCharge, monetization company Crossrider, and digital media company Matomy Media Group.

Extensive Merger and Acquisition Activity

The number of mergers and acquisitions is another excellent indicator of the vibrancy of Israel's high-tech industry. Activity has been robust, especially from United States-based multinationals seeking to acquire technology companies. European multinationals also have made acquisitions, although to a lesser extent. Asian countries, especially China, have tapped into this promising market more recently, making their first investments over the past few years.

In 2014, merger and acquisition activity in Israel's high-tech sector amounted to approximately US$5 billion from eighty-two acquisitions. The most popular trends currently driving these investments are the purchase of companies at an early stage to enjoy lower valuations and innovation cooperation, and investment in tailor-made funds such as the Israel Asian Fund (IAF) to diversify risk and enhance industrial synergies.

The table below highlights some of the acquisitions of Israeli firms by global multinationals.

Multinational Acquirer	Israeli Companies Acquired
United States	
Amazon	Annapurna Labs
Apple	Anobit, PrimeSense
Cisco	Class Data Systems, HyNEX, Seagull semiconductor, PentaCom, P-Cube Riverhead Network, Intucell, Sheer Networks, NDS Group, Infogear
Dropbox	Cloudon
eBay	Shopping.com, Fraudscience, The Gift Project
Facebook	Snaptu, Face.com, Onavo
Google	LabPixies, Waze
HP	Nur Macroprints, Mercury Interactive, Shunra
IBM	Diligent Technologies, Storwize, Worklight, Trusteer
Intel	Telmap, Omek Interactive
Microsoft	Whale Communications, Gteko, Ya Data, 3DV Systems
Monsanto	Rosetta Green
Motorola	Terayon, Bitland
SanDisk	M-Systems
Yahoo	FoxyTunes, Dapper, RayV, ClarityRay
Europe	
Orange	Orca Interactive
SAP	Top Tier Software, TopManage, A2i
Siemens	Tecnomatix, Solel Solar
Asia	
ChinaChem	Makhteshim Agan
Brightfood	Tnuva
Fosun	Alma Lasers, The Phoenix Insurance Group
Rakuten	Viber
SK Group	Camero
SingTel	Amobee

Israel's High-Tech Sectors of Excellence

The Israeli high-tech industry has taken leading positions in a variety of niche sectors. Israel's advanced expertise in diverse industries has evolved from the country's continued efforts to identify innovative solutions to problems stemming from its limited resources. Sectors of excellence include agricultural technology (agritech), clean technology (cleantech), homeland security and cyber security, and life sciences and biotechnology (biotech), among others.

A full description and listing of Israel's high-tech sectors of excellence along with the many innovators in each sector appears in Appendix 2 on page 175.

CHAPTER 6

Key Success Factors of Israel's Innovation Ecosystem

The staggering success of Israel's innovation ecosystem is underpinned by several factors. These include strong government support, a seasoned venture capital community, diversified and skilled human resources, and a highly developed technology infrastructure. The interaction of these dynamics nurtures the continued expansion of Israel's role as a leader in global technology.

Government and Public Support

Office of the Chief Scientist (OCS)

The Israeli Ministry of Economy's Office of the Chief Scientist (OCS), created in 1974 and currently headed by Avi Hasson, executes government policy in support of industrial research and development (R&D). The OCS aims to stimulate economic growth by encouraging technological innovation and entrepreneurship, leveraging Israel's scientific capabilities and providing financial support and benefits.

MATIMOP Israeli Industry Center for R&D

The OCS has stimulated innovation through several programs operating under the umbrella of the MATIMOP agency, which coordinates international R&D programs. Bilateral R&D cooperation agreements have been signed with countries in North America (the

United States and Canada), Europe (Germany, France, The Netherlands, Russia, Switzerland, and the United Kingdom), and Asia (China, India, and Taiwan).

A number of binational R&D funds have also been established with North American and Asian countries. These include the binational fund with the United States, the Israel–US Binational Industrial R&D Foundation (BIRD); with Canada, the Canada–Israel Industrial R&D Foundation (CIRDF); with Singapore, the Singapore–Israel Industrial R&D Foundation (SIIRD); and with South Korea, the Korea–Israel Industrial Foundation (KORIL).

Additional multilateral agreements have been made with EUREKA, the largest global program promoting industrial innovation; Galileo, a European program for satellite navigation and positioning systems; ERA-NET, for European cooperation in research activities; and Horizon 2020/ISERD, for European Union R&D programs.

Creation of Israel's Venture Capital Industry

Yozma Program

Israel's venture capital industry originated on the back of private initiatives. One of the pioneer venture capital (VC) firms, Athena Venture Partners, was founded in 1985 by Israel's former Air Force Chief of Staff Major General Dan Tolkowsky, Dr. Gideon Tolkowsky, and Frederik Adler, a veteran of the American VC industry. Veritas Venture Capital Management, another pioneering firm, was established in 1990 by Dr. Gideon Tolkowsky and Yadin Kaufmann.

However, the political willingness of Israel's government to create a sustainable local VC industry to finance early-stage innovation is what ultimately enabled Israel to gain stature as a global technology center. In 1993, the OCS established the Yozma Program (*yozma* means "initiative and entrepreneurship" in Hebrew). At the same time, it created a US$100 million government-owned VC company that invested US$80 million in ten private early-stage VC funds (US$8 million per fund), and another US$20 million in direct investments in high-tech firms.

In order to receive government funding, each VC fund had to comply with several conditions. The first was to create a new VC firm that was not held by incumbent financial institutions to ensure greater competition in the market. The newly created VC also needed to have a local Israeli management team as well as established investors, such as foreign limited partners and local financial institutions. The Israeli government's investment attracted an additional US$150 million of domestic and international investments.

The success of this scheme stimulated the development of a professional VC industry in Israel. The involvement of experienced foreign limited partners dramatically increased the learning curve of local participants in the field.

In only one year, from 1992 to 1993, Israeli VC fundraising increased to US$162 million from US$27 million. Since this landmark growth period of the 1990s, Israel's VC industry has experienced strong growth both in terms of capital raised and formation of VC funds. With the exception of 2009 and 2010, the Israeli VC industry raised some US$1 billion each year on average. [37]

The Yozma program was voluntarily phased out in 1998, after five years of operation, when private investors leveraged their profits by purchasing the Israeli government's shares in the funds. By then, it had enabled Israel's nascent VC industry to thrive in two very significant ways. First, it provided risk sharing and an attractive investment incentive scheme for investors. Second, it strongly encouraged entrepreneurs in the Israeli technology ecosystem to create innovative start-ups with the comfort of a wide availability of early-stage funding.

---

[37] IVC Research Center.

Yozma Funds: The Backbone of Israel's Venture Capital Industry

Name	Vintage	Original size (US$mn)	Foreign LP	Country
Yozma	1993	20	None	Israel
Gemini	1993	25	Advent	USA
Inventech	1993	20	Van Leer Group	Netherlands
JVP	1993	20	Oxton	USA
Polaris (Pitango)	1993	20	CMS	USA
Star	1993	20	TVM, Singapore Tech	Germany, Singapore
Walden	1993	25	Walden International	USA
Eurofund	1994	20	Daimler-Benz, DEG	Germany
Nitzanim	1994	20	AVX, Kyocera	USA, Japan
Medica	1995	20	MVP	USA
Vertex	1996	20	Vertex International	Singapore

Source: Yozma

Incubator Program

The OCS also created an Incubator Program in 1991, which remains in operation today. The program is designed to leverage the extensive knowledge base of the hundreds of thousands of engineers and scientists who immigrated to Israel from the former USSR between 1986 and 2006.

The OCS started by creating six incubators to encourage seed and early-stage technology development through entrepreneurship. Today, there are twenty-seven privately-owned incubators operating under an eight-year license from the OCS through a competitive tender. Each incubator hosts eight start-ups on average for a two-year period. The average individual budget of US$500,000 per start-up is covered by OCS grants of up to 85% of the total budget, with 15% funded by private investors. The OCS recoups its investment through royalties of 3%-5% of the total revenues of successful ventures.

Like the Yozma Program, the Incubator Program has played a critical role in stimulating and enhancing Israel's innovation ecosystem.

Technology Clusters

One of the strengths of Israel's public and government policies was to encourage the emergence of technology clusters in various parts of the country. Today, Israel has some twenty-seven high-tech parks, spread widely over its territory, covering various fields of innovation. This diversification provides a unique opportunity for the local population to participate in its regional innovation ecosystem. It also enables the emergence of regional competence centers based on the attributes of various Israeli territories. Beersheba, for example, has built capabilities and expertise in the renewable energy sector, especially in solar energy, due to its proximity to the Negev Desert.

Northern Israel: Haifa, Yokneam, Caesarea

Haifa, one of Israel's largest technology clusters, is home to several technology business parks. Matam Park hosts numerous start-ups, large high-tech companies, and the R&D centers of IBM, Intel, Yahoo, Google, NDS Group, Elbit Systems, and others. Haifa also has a dedicated life sciences park, which hosts life sciences companies and research institutes, such as the Rappaport Family Institute for Research in the Medical Sciences and the Rambam Health Care Campus. The city also hosts different branches of the Technion Institute of Technology and Haifa University.

Yokneam's industrial park hosts more than 100 leading technology firms. Companies represented include Intel, Panasonic, Given Imaging, Mellanox Technology, Marvell Technology, and Lumenis. The nearby major technology cluster at Caesarea is another state-of-the-art high-tech center hosting more than 170 companies.

Central Israel: Herzliya, Tel Aviv, Ra'anana, Jerusalem

The largest technology clusters are located in central Israel. The most extensive of these is in Herzliya, which hosts several high-tech parks. These are home to global technology giants such as Apple, Siemens, CA, Microsoft, and RSA and to leading Israeli technology groups such as Verint, Matrix, and numerous start-ups. The Interdisciplinary Center (IDC) Herzliya, a leading Israeli university, also participates actively in the fertile local ecosystem.

Microsoft building in Herzliya, Israel's flagship high-tech park

Tel Aviv is home to numerous global technology companies such as Google and Facebook. It also hosts numerous start-ups. Tel Aviv promotes itself to Israeli and international entrepreneurs by offering various incentives, such as subsidies for entrepreneurs, municipal tax incentive plans, and co-working space facilities. The presence of Tel Aviv University and Bar-Ilan University enriches this ecosystem.

The high-tech park in nearby Ra'anana is also home to a number of innovative companies, including Israeli global players Amdocs and Nice Systems, as well as Sarana Ventures' incubator featuring Teva and Philips.

Jerusalem supports the Israeli ecosystem with several business parks hosting such global companies as IBM and BrightSource Energy. It also hosts leading Israeli conditional access firm NDS, bought by Cisco, and numerous innovative start-ups. The Hebrew University and Jerusalem College of Technology enhance this local innovation ecosystem.

Southern Israel: Beersheba

Beersheba, located in the Negev Desert, has been a key technology cluster for many years. The city boasts several high-tech parks, including a new advanced Technology Park, which is the State of Israel's National Cyber Security Headquarters. Companies located in Beersheba include Deutsche Telecom, EMC, Ness, DbMotion, Oracle, Elbit, Lockheed Martin, IBM, RAD, and Audiocodes. The Ben-Gurion University of the Negev contributes to the innovative ecosystem and hosts the leading medical institution in the area, Soroka Medical Center.

Israeli and Foreign Venture Capital

The number of Israeli VC funds has increased dramatically to more than 115 in 2015 from about a dozen in the 1990s. Israeli players, such as Pitango, Carmel, JVP, and Gemini, dominate the country's VC market. Initial investors in the sector were Israeli financial institutions followed by US institutions. Today, these VCs have diversified their investor base with European and Asian corporate and institutional investors.

Foreign firms, lured by Israel's blossoming high-tech innovation ecosystems, command nearly 50% of the market. Foreign funds operate in several ways: through a branch office, directly from abroad, or through the corporate venture capital format (for example, Intel Capital).

The majority of foreign investors are United States-based and heavily weighted toward Silicon Valley, such as Sequoia, Benchmark Capital, and Blumberg Capital. Several European VC firms also operate in Israel, including Kreos Capital and Alta Berkeley. Asian firms have developed a stronger presence recently and are growing rapidly. The most notable of these are Li Ka-shing-controlled Horizon Ventures, Samsung Ventures, and the Israel Asian Fund.

The United States still provides the largest source of funding for Israeli VCs. Firms from the United States represent about 70% of the total pool of limited partners (LPs) in the Israeli VC market. The remaining 30% of LPs come from Europe and Asia.

The dynamic is changing rapidly, however, with the economic rise of Asia (and predominantly China). Within five to ten years, it is estimated[38] that both Asia and the United States will be the leading suppliers of capital to Israel, with about 40% each of the LPs, while Israel and Europe will have some 10% each.

Skilled Human Resources

Immigration

Israel is a country of immigration. Since its inception in 1948, more than 3 million immigrants have settled in Israel, which now has a population of about 8 million. Among these, more than 1 million come from the former Soviet Union, about 150,000 from North America, 80,000 from France, 70,000 from Argentina, 35,000 from the United Kingdom, and 20,000 from South Africa.[39]

Most Israeli immigrants have a strong educational background and many of them, particularly those from the former Soviet Union,

---

[38]   E&Y and IVC Research Center.
[39]   Israel Central Bureau of Statistics.

hold advanced degrees in engineering, mathematics, physics, and life sciences. These hundreds of thousands of gifted scientists now form the backbone of Israel's high-tech industry, especially in the field of research and development. The remaining immigrants from all over the world boost the ecosystem by employing their diverse and complementary skills, as well as their connections around the world.

Entrepreneurial Culture

At the inception of the State of Israel in 1948, the country lacked natural resources and had limited infrastructure. New immigrants, many of whom came for ideological reasons, found themselves building the country from scratch in every area, from roads to railways, communications to utilities, and administration to provisions for basic needs. Israel's pioneers quickly developed the entrepreneurial skills needed to survive and eventually thrive.

The spirit of fighting to survive in a hostile environment forged the mentality of the local people. Israeli creativity in innovation and technology finds its roots in these early years when success depended on identifying imaginative solutions at the lowest possible cost. The national slogan became *"ein Breira"* – "there is no choice, success must be achieved". Neither the State nor its citizens had abundant financial resources. A melting pot of individuals, mentalities, backgrounds, languages, and education gave birth to a new society with a mantra of entrepreneurship as the basis of salvation. This way of thinking empowered Israeli citizens to change the world with their ideas and workforce.

Today in the technology arena, the Israeli approach of never giving up has underpinned the country's accomplishments in global innovation. Thinking out of the box is not a catch-phrase for Israelis. It is ingrained in their mental fiber as the only way to approach problems and achieve results in a creative, cost-effective way.

Experienced Second-Time Entrepreneurs

Israelis are born entrepreneurs, because they had no choice but to take risk launching different ventures. The country had no companies to speak of in 1948, and employment in administration was limited. The only way to survive was to create an enterprise.

The development of the global IT sector in the 1960s and 1970s enabled many Israelis with scientific backgrounds to apply their creativity in technology-related services. They created the first start-ups of their time, often under the leadership of one group. The Rad Group, for instance, empowered more than fifty managers, who in turn created more than 100 companies of their own.

The "second-time" entrepreneur effect is another key factor in nurturing the success of Israel's innovation ecosystem. The majority of founders of Israeli companies who successfully realized an exit through either mergers and acquisitions or initial public offerings have chosen to start new ventures. Because the founders are usually part of "Angel" investments or mentorship programs, their knowledge is shared with a larger community, thereby further multiplying the potential number of new ventures. It is not unusual to find serial entrepreneurs who have created three, four, five or even more start-ups during their lifetimes.

University Computer Science Students

A regular stream of well-educated computer science graduates entering the workforce contributes to the dynamism of Israel's innovation. Computer science is one of the most popular majors for university students both at the undergraduate and graduate levels. It is considered prestigious to attend an elite university like the Technion, Hebrew University, Tel Aviv University, or Ben-Gurion University in the field of computer science, because many graduates know they will find well-paid employment in the high-tech sector. The desire to create independent start-ups later in their careers strongly motivates these employees to succeed.

Foreign Technology Firms

International foreign companies have been active in Israel since the 1960s and 1970s when IBM and Intel pioneered R&D activity in the country. In the early 2000s, many more American corporations – HP,

General Electric, Motorola, Microsoft, Google, Yahoo, Cisco, Apple, and Facebook, to name a few – established a footprint in the country. At the same time, European firms, such as Siemens, Alcatel-Lucent, Dassault Systems, Philips, and SAP, started operations. Asian corporations too, including Samsung, LG, ZTE, Huawei, and Toshiba, have developed a presence in the country over the past five years.

Foreign firms both learn from Israel's innovation market and nurture it through the sharing of best practices. Extensive foreign entry into the local market has stimulated the ecosystem by providing ongoing opportunities for emulation in terms of R&D capabilities and a productive, competitive environment.

Technology Infrastructure

Leading Academic and Research Institutions

As outlined in Chapter 4, academic and research institutions have developed the infrastructure for technology innovation. State-of-the-art laboratories and technology transfer companies help bridge the gap between fundamental research and product commercialization.

Defense Companies

Israel's largest defense companies, such as Israel Aerospace Industries (IAI), Elbit Systems, and Rafael, contribute greatly to the development of Israel's innovation ecosystem. Their advanced R&D capabilities allow them to create breakthrough technology that can adapt to multiple uses, and require a high-standard working methodology and highly trained staff. Many technologies produced for the military market have been used in civilian applications. For example, the unmanned air vehicle (UAV), created under the leadership of Dr. David Harari at IAI, was designed originally for military purposes. Now it is used for diverse civilian applications in several fields, including agriculture for crop observation, mines for safety monitoring, construction sites, infrastructure inspection, wildlife research, environmental monitoring, and search and rescue.

Corporate Research and Development

Corporate R&D is another major component of the technology infrastructure of Israel's innovation ecosystem. Leading Israeli firms, like TEVA in the pharmaceutical sector, as well as major international corporations have developed centers in Israel. Multinationals with R&D centers in Israel include 3M, AOL, Apple, Applied Materials, AT&T, AutoDesk, Berkshire-Hathaway, BMC Software, Broadcom, Cisco, Dassault Systems, Dell, eBay, EMC, Facebook, General Electric, General Motors, Google, HP, Huawei, IBM, Intel, Johnson & Johnson, LG, McAfee, Microsoft, Motorola, Oracle, PayPal, Philips, Polycom, Qualcomm, Samsung, SanDisk, SAP, Siemens, Texas Instruments, Toshiba, Xerox, and Yahoo.

Corporate Spin-offs

Corporate spin-offs also contribute to the technology infrastructure of Israel's innovation ecosystem as their activity is voluntarily, for diverse reasons, lodged in a separate vehicle from their existing shareholders. The spin-off often focuses entirely on a particular market or technology.

Incubators

As of 2015, twenty-seven OCS-licensed incubators operate in Israel. They contribute significantly to Israel's technology infrastructure ecosystems by accelerating the business development of start-up firms. Key incubators include Abital Pharma Pipelines (life sciences); Alon-MedTech Ventures (health technologies); Explore; Dream Discover (Internet, mobile, new media); Incentive Incubator (medical devices and software); Incubit Technology Ventures (medical devices, ICT, cyber, electro-optics); JVP Cyber Labs (cyber security and enterprise software); JVP Media Studio (new media and enterprise software); Kinrot Ventures (clean technologies); Trendlines Medical (medical technology); Trendlines Agtech (agritech); The Time Innovation (new media); Van Leer Xenia (medical devices, advanced materials, industrial applications and IT); NGT (life sciences); Nielsen

Innovate Fund (Internet, communication, mobile, telecom, media); Rad BioMed Incubator (life sciences and medical devices); Terralab Ventures (cleantech); FutuRx (biotechnology); and Sarana Ventures (medical technologies).

There are also private incubators, which operate outside the framework of the OCS incubator programs. Large global high-tech companies often sponsor autonomous private incubators, such as the Nazareth Business Incubator Center, supported by Cisco Systems.

Accelerators

Accelerators help to speed up the growth rate of start-ups. Their programs can take a number of forms, with variations in length of program from a few weeks to a year, and in amount of financial and mentoring support provided, Key accelerator programs in Israel include: Microsoft Azur Accelerator (Microsoft); Nautilus (AOL); nazTech Accelerator (Cisco); Google Campus Accelerator (Google); IBM Alpha Zone Accelerator (IBM); FutuRX (Johnson & Johnson); BluePrint Accelerator (PayPal); Samsung Accelerator (Samsung); Yahoo Accelerator (Yahoo); Citibank Accelerator (Citibank); IDC Elevator; DreamIT Ventures; and Gvahim.

CHAPTER 7

Israel on the Roadmap of Asian Investors

Israel has been on the radar of United States technology giants for decades. Numerous multinational corporations have successfully developed strong operations in the country, either through direct activity, R&D centers, investment in funds or M&A activity.

The predominance of the United States, and to a lesser extent Europe, has, however, been seriously challenged over the past few years by Asia. Although Asian countries, especially Japan and Singapore, initiated investments in Israel before 2010, mainly through investment funds, the volume was marginal compared with current levels of engagement.

The rapid shift in appetite of Asian investors, now dominated by China, coincides with two events that dramatically changed the face of the Israeli economy. First, the nomination of Stanley Fisher as Governor of the Bank of Israel coincided with an unprecedented era of economic growth and stability. Second, the discovery of large offshore gas fields in Israeli territory positioned the country as a key global energy producer. These favorable market conditions encouraged the first large, groundbreaking transactions that subsequently stimulated further investment and business cooperation between Israel and Asia.

The Fisher Effect

Governor of the Bank of Israel

Stanley Fisher, one of the world's leading economists and central bankers, was appointed Governor of the Bank of Israel in May 2005.

Born in what is now Zambia, he holds dual United States and Israeli citizenship and was educated at the Massachusetts Institute of Technology (MIT) and the London School of Economics. Previously, he held several senior positions, including Vice Chairman of Citigroup, First Deputy Managing Director of the International Monetary Fund (IMF), and Vice President for Development Economics and Chief Economist at the World Bank. Interestingly, he was the thesis adviser of both former US Federal Reserve Chairman Ben Bernanke and European Central Bank Governor Mario Draghi at MIT. Fisher was rated among the world's top six central bankers in 2012 by *Global Finance* magazine and named Central Bank Governor of the Year by *Euromoney* magazine in October 2010.

Fisher had a strong influence on Israel's economy during his tenure as Governor of the Bank of Israel. He was the first central banker to cut interest rates in 2008 at the start of the financial crisis, and the first to raise them again following signs of recovery. Under his leadership, the Bank of Israel purchased large quantities of US dollars with Israeli shekels to prevent a sudden increase in value of the Israeli currency that could have harmed Israel's competitiveness in global markets. He also advanced the Bank of Israel's autonomy and governance by promoting a new Bank of Israel Law (2010) that ensured the bank's independence. He created a Monetary Policy Committee, composed of six leading economic and financial experts, to decide consensually the bank's monetary policy. These experts included Barry Topf, Senior Advisor to Stanley Fisher, who had previously been responsible for managing Israel's foreign exchange reserves.

Fisher left the Bank of Israel in June 2013 and now is Vice Chairman of the Board of Governors of the United States Federal Reserve System. In his farewell speech to Fisher, Israeli Prime Minister Benjamin Netanyahu praised the critical role he had played in the economic growth of the State of Israel and the achievements of the Israeli economy.

Extensive travels to Asia positioned Fisher well to advocate for Israel and its economy with Asian investors. The country's positive economic fundamentals, demonstrated by its stability during the financial crisis that began in 2008, and its unique innovation capacity, have since attracted the attention of Asian investors. Within a short span of time Israel has been earmarked as a key center for Asian innovation investment.

Credit Ratings

Stanley Fisher's arrival at the helm of the Bank of Israel in 2005 contributed to the global attractiveness of Israel's economy. In the ten years prior, major credit agencies such as Standard & Poor's (S&P), Moody's and Fitch had rated Israel's long-term debt with an average "A-/A3". As structural reforms began under Fisher's leadership and successful policies were implemented to protect the country from the global economic crisis, S&P upgraded Israel's long-term credit rating to an "A" in 2007 and later, boosted by the country's buoyant financial prospects, to "A+". Around the same time, Moody's and Fitch also upgraded their ratings to "A1" and "A", respectively. Today, Israel's long-term debt is rated "A+" by S&P, "A1" by Moody's, and "A" by Fitch.

This vote of confidence in the Israeli economy has helped tremendously in attracting foreign investors, including Asian participants that had historically been highly sensitive to the stability and solidity of Israel's economic fundamentals.

Admittance to International Bodies

Already in 2003, Israel's financial stability and sophistication were recognized when it was invited to join the Bank for International Settlements (BIS). The BIS is an international organization of central banks that fosters international monetary and financial cooperation. It is known as the "Central Bank for the world's central banks".

In 2010, Israel joined the Organization for Economic Cooperation and Development (OECD) which is composed of thirty-four leading economic powers[40]. This important milestone for Israel was made possible through the guidance and leadership of Stanley Fisher. Israel's integration into the global economic landscape has strengthened investors' confidence in the economy and encouraged more investments, especially from Asia. Additionally, Israel's commitment to compliance with advanced international standards has enabled it to promote necessary socioeconomic reforms.

Israel's entry into the OECD has entitled it to receive funding from international investment funds that invest only in OECD countries. This factor has dramatically transformed Israel's dynamics. The adhesion of Israel to the OECD provides international recognition of the country's ability to join this "prestigious club" of powerful economic nations and reflects the increasing attractiveness of its unique ecosystem.

MSCI Index

The Morgan Stanley Capital International (MSCI) World Index is the leading stock market index of more than 1,600 stocks from around the world. It achieved an annual return of 5.5% in 2014. The index includes a basket of stocks from the developed markets; emerging countries' stocks are excluded from this key index.

In May 2010, the MSCI reclassified Israel as a developed market. With this move, Israel joined other developed countries including Australia, Austria, Belgium, Canada, Denmark, Finland, France, Germany, Hong Kong/China, Ireland, Italy, Japan, Netherlands, New Zealand, Norway, Portugal, Singapore, Spain, Sweden, Switzerland, the United Kingdom, and the United States.

The change in status from emerging to developed market has acted as a spur to foreign investors, including Asian participants. The

---

[40]   Austria, Australia, Belgium, Canada, Chile, Czech Republic, Denmark, Estonia, Finland, France, Germany, Greece, Hungary, Iceland, Ireland, Italy, Japan, Korea, Luxembourg, Mexico, Netherlands, New Zealand, Norway, Poland, Portugal, Slovak Republic, Slovenia, Spain, Sweden, Switzerland, Turkey, United Kingdom and the United States.

reclassification was seen as another vote of confidence in the Israeli economy.

The Israeli Shekel in Global Foreign Exchange

The Continuous Linked Settlement System (CLS) was established in 2002 by a group of foreign banks to reduce risk in settling foreign currency transactions. It operates as an international clearing house for foreign currency conversion. CLS is owned by more than seventy financial institutions worldwide and is supervised by the US Federal Reserve Bank in coordination with representatives from the countries whose currencies trade on the CLS platform. CLS carries out settlement and conversion activities from currency to currency simultaneously. Today, the CLS platform provides settlement services for seventeen global currencies.

In May 2008, under the leadership of Stanley Fisher, the Israeli shekel joined CLS, significantly reducing the settlement risks involved in any transaction between local and international payments. The inclusion of the shekel among the world's leading currencies on the CLS platform enabled its full convertibility and liquidity internationally and helped Israel implement reform in its payment and settlement systems.

The Gas Effect

Gas Discovery Overview

The Delek Group, led by Yizhak Tshuva, is Israel's dominant integrated energy company. Its pioneering activity in natural gas exploration and production is transforming the Eastern Mediterranean's Levant Basin into one of the energy industry's most promising emerging regions. Together with its American-based partner, Noble Energy, they have been active in the eastern Mediterranean Sea since 1998. The Delek Group discovered the Yam Tethys gas field with the Noa and Mari B reservoirs in 1999 and 2000, respectively. These reservoirs are located 25 kilometers off the coast of Israel. On the back of this discovery, Noble

Energy and the Delek Group have supplied Israel with natural gas since 2004.

The same partners discovered the Tamar gas field in 2009. Tamar was the world's largest natural gas discovery that year, with 10 trillion cubic feet of natural gas and the capacity to meet Israel's market demand for twenty years. In April 2013, the Tamar gas field began supplying natural gas to the Israeli market.

Noble Energy and the Delek Group discovered the Leviathan gas field in 2010. With 19 trillion cubic feet of natural gas, this became the world's largest discovery of natural gas in deep water in the last decade. Noble Energy, the Delek Group, and other partners also have discovered a number of smaller fields[41].

The discovery of natural gas fields, as well as new oil and gas capabilities, which were virtually non-existent before the 2000s, has enabled Israel to join the select club of nations that are global energy providers. Israel's offshore gas discovery is expected to have a tremendous impact on the local economy and its ability to attract foreign investment, especially from Asian countries.

---

[41]   These include the Karish, Tanin, Dolphin, Ruth, David, Keren Avia, Eran and Alon fields.

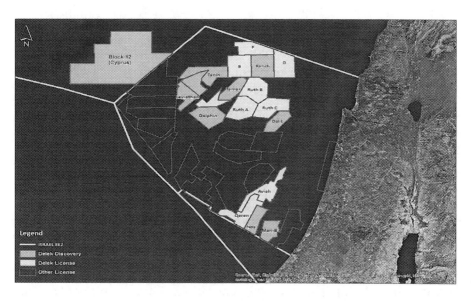

Israel's offshore gas assets

Delek's offshore natural gas platform in the Mediterranean Sea

Increasing Global Competitiveness

The discovery and exploitation of gas in Israel is a landmark event that will dramatically change the face of the local economy. It will also attract more foreign investment, especially from Asia, looking for ways to secure energy and commodity resources.

Israel's offshore natural gas discovery has enabled it to achieve a greater degree of energy independence, which is today of the highest strategic importance, given the global race to secure energy and commodities. Israel's transition to natural gas is expected both to increase the state's energy management efficiency and reduce its energy cost by approximately 10%[42]. All industries, from low-medium-high-tech to traditional industries such as chemicals, plastics, textiles, paper, food, as well as agriculture will benefit.

The gas discovery will strengthen the finances of the State of Israel. Bank Leumi estimates the Israeli Tax Authority will collect US$30 billion in gas royalties on licenses, US$30 billion in corporate taxes, and US$70-80 billion in excess profits tax during the period of operations (2013-2040) of the discovered reservoirs. The gas royalties and corporate taxes are expected to be incorporated into the government budget while the excess profit tax will be invested into an Israeli sovereign wealth fund. Over the long term, these moves should reduce the state's debt level, help maintain a balanced budget and increase investment and spending for the Israeli economy. Improved fundamentals will lead to further infrastructure enhancements and the creation of new industries and sectors of excellence, which will heighten Israel's global competitiveness.

Israel's sovereign wealth fund will invest entirely in foreign assets, as part of a strategy to fight "Dutch Disease", a term deriving from the Netherlands' experience when a natural gas discovery in the 1960s dramatically increased currency inflows and led to a currency appreciation that harmed the competitiveness of its manufacturing sector.

---

[42]   Source: Bank Leumi.

China's Landmark Deals

Asian countries, though historically not as active as their American and European counterparts in the Israeli economy, have come to understand the value proposition of Israel. Many Asian countries have adopted strategic roadmaps to invest and increase their foothold in the unique Israeli ecosystem. The expected inflows of money and investment will create numerous business opportunities.

The favorable economic environment of Israel and its unique ecosystem were identified at an early stage by outstanding business leaders such as Hong Kong's Li Ka-shing. These visionary leaders have played a commanding role in opening the doors to Israel–Asian investment and business cooperation. Since the 2000s, Li Ka-shing's group of companies has participated actively in Israel's growth. This early activity has encouraged other large Chinese groups, such as ChemChina and Bright Food, to enter the Israeli market through landmark gate-opening transactions.

China's strong appetite for the Israeli market, with landmark deals outlined below, has provided a strong catalyst for other Asian countries to intensify their investment and foothold in the country.

Li Ka-shing and Hutchinson Whampoa

Outstanding Individual Track Record

Li Ka-shing is Asia's wealthiest individual, with an estimated net worth of US$33.5 billion in 2015[43]. Since the 1950s, he has built a global diversified business group with a wide range of activities. His stable of companies includes leading conglomerate Hutchison Whampoa, whose activities encompass the world's largest operator of container port facilities; the world's largest health and beauty retailers Superdrug (United Kingdom) and Marionnaud (France); leading retail distributors Park n Shop and Fortress electrical appliances (Hong Kong); and utility company Hong Kong Electric. Li Ka-shing's CK Hutchison Holdings Limited controls twelve listed companies

---

[43]   Source: *Forbes.*

and operates in more than fifty-five countries with approximately 260,000 employees.

Love Affair with Israel

Li Ka-shing has always admired Israel and the Jewish entrepreneurial spirit. His success story in Israel has been a strong gate opener for Asian investors who wish to follow closely in the steps of Asia's wealthiest businessman. In addition to Li, another of the world's wealthiest individuals and most respected entrepreneurs and business gurus, the American Warren Buffet, has also been a strong advocate of the Israeli economy. Buffet's Berkshire Hathaway made its first acquisition abroad in Israel in 2006 when he purchased an 80% stake in Iscar Metalworking, a manufacturer of metal cutting tools, for US$4 billion.

The Hutchison Whampoa Group entered the Israeli market in 1999, when Hutchison Telecom won Israel's third mobile network operating license alongside Israeli firms Elbit and Eurocom. Together they launched Israeli firm Partner Communication, which continues to operate under the Orange[44] brand name.

In 2001, Partner Communication listed on the NASDAQ and London Stock Exchange, and Hutchison Telecom became the largest controlling shareholder with 51.6%. In 2009, Hutchison Telecom sold its controlling holding in Partner Communication to Israeli businessman Ilan Ben-Dov[45] for approximately US$1.4 billion.

Enthusiastic about the local innovation ecosystem and Israel's market opportunities, Li Ka-shing gave a mandate to Hutchison Whampoa's Israeli team, headed by Dan Eldar, to suggest other venues for investments. They proposed the water sector, which Li Ka-shing selected as one of Israel's core fields of excellence. Accordingly,

---

[44]   Hutchison Whampoa sold the Orange Group to German conglomerate Mannesmann in October 1999 for US$33 billion. British group Vodafone acquired Mannesmann in February 2000 and sold Orange to French group France Telecom for EUR40 billion in August 2000.

[45]   Partner Communication currently is held by Haim Saban Private Equity firm, Ilan Ben-Dov and other investors, and the public.

Hutchison Whampoa established a new operation in the water segment in 2008, with capabilities in desalination, water treatment, wastewater treatment, water reuse and water technologies. Israel quickly became Hutchison Whampoa's global center for water expertise.

In 2011, Hutchison Water successfully closed its first landmark transaction in the field. Hutchison Water International and its partner, Israeli company IDE Technologies, won the tender to build the world's largest seawater reverse osmosis desalination plant at Sorek in Israel. The Sorek plant now provides more than 150 million m^3 of clean water supply to more than 1.5 million Israeli citizens.

In 2012, Hutchison Whampoa's investment in the Israeli water industry strengthened as the firm won the Kinrot Ventures technological incubator license granted by the OCS. This incubator specializes in water and cleantech seed investments. Hutchison Whampoa committed to investing at least US$25 million over the following eight years.

Li Ka-shing's venture capital arm, Horizon Ventures, has been one of the most active VCs in Israel for the past few years, acknowledged as the most active investor in 2013 and second-most active in 2012. Horizon Ventures focuses on disruptive technologies. It has invested in many of Israel's leading innovative companies, such as Waze, FeeX, Tipa, and Kaiima.

China National Chemical Corporation and Makhteshim Agan

In October 2011, China National Chemicals Corporation (ChemChina) made a historic acquisition of a 60% controlling shareholding stake in Makhteshim Agan, a leading Israeli manufacturer of crop protection products, including herbicides, insecticides, and fungicides. ChemChina paid US$2.4 billion, of which US$1.2 billion went to public shareholders and US$1.1 billion to Koor Industries (IDB Group), its former controlling shareholder. After the completion of this landmark transaction, China's largest investment in Israel to date, Israeli company Koor Industries remained the second-largest shareholder with a 40% stake. In 2014, Makhteshim Agan was renamed ADAMA. Under the leadership of its new parent company it has continued to command the global fertilizer and crop protection market.

The size and strategic importance of this transaction set a milestone for Israeli–Chinese business relations. Both Israel and China realized the tremendous potential of greater business cooperation, with Israel providing innovation and technology capability and China bringing the necessary financial resources and market depth required for a company to compete on a global level. This deal served as a business case for Chinese investors wishing to fully realize the advantages of investing in the Israeli ecosystem. It also provided a successful example to replicate.

Following this landmark transaction, China's investments in Israel increased dramatically, and many other deals came to fruition.

Bright Food and Tnuva

In March 2015, China's Bright Food completed the acquisition of a 56% controlling stake in Tnuva Food Industries for approximately US$1.2 billion from financial fund Apax Partners and an additional 21% from another investor, Mivtach Shamir, for approximately US$275 million. The *kibbutzim* (collective farms) movement remains a shareholder of Tnuva with 23%.

Tnuva is Israel's largest cooperative with more than 620 members, mainly *kibbutzim* and *moshavim* (agricultural communities), specializing in milk and dairy products. It accounts for more than 70% of the local market. Tnuva's advanced technologies and production capabilities captured the interest of Bright Food.

This landmark deal between a Chinese investor and Israel's primary milk and dairy producer is another example of the industrial cooperation that can benefit both countries. With its financial capabilities and the size of its local market, China brings the necessary drivers for sustained long-term growth, while Israel with its cutting-edge technology and innovation provides its unique state-of-the-art production capabilities. The deal has sent a strong signal to Chinese and Asian investors that Israel has developed a unique economic ecosystem that can optimize synergies between their economies.

Through its buoyant economy, its unique ecosystem and its newly found gas resources, Israel has succeeded in attracting the "leader of opinion" deal makers. These individuals, in turn, have convinced an ever-growing number of Asian investors that Israel should be at the center of their global investment strategy.

PART THREE

ASIA WORLD BUSINESS CENTER

CHAPTER 8

Asian Activity in Israel

Japan and Singapore were the first Asian countries to engage in substantial business activity with Israel. During the 1970s-1990s, Japan exported major automotive brands, including Subaru, Daihatsu, Suzuki, Honda, Mazda, Toyota and Nissan, to Israel. In the mid-1990s, it also participated in the financing of Israel's venture capital industry, with Kyocera, JAFCO, Hitachi, NTT, and Mitsubishi, alongside leading Singaporean investor Vertex.

More recently, China has become the most active investor by far, spurred on by Hutchison Telecom's investment in Partner Communication/Orange mobile telephone operations in 1999. Bilateral trade between Israel and China has blossomed since the 2000s with an estimated US$9 billion exchanged in 2014. According to the Ministry of Economy, China's investment in Israel has increased dramatically in the past few years. China ranked first in 2014 in the number of joint projects with the Israeli Office of the Chief Scientist (OCS). In 2013, it ranked second, while in 2012 it was not even a contender in the ranking[46].

China in Israel

The diversity and depth of China's business activity in Israel clearly demonstrate the evolution of the Chinese–Israeli relationship. We highlight below some of the more outstanding commercial transactions and investments between Chinese and Israeli entities to

---

[46] Reuters.

illustrate the growing commitment between the two countries. We also outline landmark transactions and activities undertaken with other Asian countries, including Singapore, South Korea, India, Japan, and Taiwan.

Direct Investments

Hutchison Whampoa

Besides the various Hutchison Whampoa Group activities in Israel outlined earlier, Li Ka-shing is very active in Israel through the venture capital fund Horizon Ventures. Horizon Ventures has made more than twenty-four investments in Israel over the past few years, outnumbering its twenty-one investments in the United States.

Its portfolio includes the following companies (listed in alphabetical order).

Company	Areas of Activity
Accelta	Provides revolutionary solutions for affordable, high-quality mass manufacturing of stem cells for research and clinical use
Aniways	Offers intelligence platform to provide lively, appealing, interactive content to users, thereby driving revenue for customers
Corephotonics	Improves smartphone photography by bridging quality gap between compact cameras and digital still cameras using cutting-edge computational photography technology
Cortica	Automatically extracts the core concepts in images and videos and maps these to keywords and textual taxonomies
Crosswise	Cross-device identification technology helps customers target audience across multiple devices

Company	Areas of Activity
Everything.me	Changes relationship between people and their phones by creating phones that anticipate users' needs
FeeX	Delivers clear information about fees related to individual investments, such as retirement plans
Ginger Software	Develops language-enhancement technology to improve English-language expression
Hola	Allows access to global websites from anywhere in world
Kaiima	Leading genetics and breeding technology provider for sustainable agriculture; provides new varieties of key crops with better yields
Magistro	Transforms plain video and photos into professional-looking movies
Medical Cancer Screening	Innovative personalized cancer solutions for physicians and healthcare centers screen for elevated probability of disease, helping reduce morbidity and mortality
Meekan	Develops flexible time scheduling engine that matches calendars, pinpointing times to meet, schedule services, reserve hotels, etc
MeMed	Produces disruptive technology decoding immune responses to guide treatment of infectious diseases, empowering physicians to reduce antibiotic misuse and address antibiotic-resistant bacteria
Meteo Logic	Offers innovative forecasting solutions based on big data and smart machine learning approach that is simple to implement and accurate
Mishor 3D	3D technology projects safety, navigation and media information on vehicle windshields while keeping driver's focus on road

Company	Areas of Activity
NanoSpun	Develops key building block for enhanced bioprocess and controlled-release solutions
Nipendo	Its platform allows businesses to connect and integrate existing systems and processes, with no custom coding or mapping required
Ovavo	Developer of mobile app analytics solutions; acquired by Facebook. Became Facebook's first development center in Israel
Shine Security	Antivirus technology performs behavioral monitoring on end devices to stop newly emerging threats
Stevie	Turns social web into television, creating personal, monetizable entertainment across platforms
Tipa	Develops 100% biodegradable flexible packing solutions for food and beverage companies; packaging recycles back to nature quickly
Waze	Social GPS application for smartphone that provides real-time information about traffic generated by users. Won Best Overall Mobile App award at 2013 Mobile World Congress. Acquired by Google in 2013 for US$1.3 bn. Horizon Ventures, with 11% stake, made exit estimated at US$143 mn.
Wibbiz	Uses advanced text analysis and artificial intelligence to create video summaries from text articles

Continuing a long tradition of philanthropic activity, Li Ka-shing donated US$130 million to the Technion Institute of Technology in 2013 through The Li Ka Shing Foundation. The donation was funded by capital gains realized from the Waze exit. Funds will be used to develop the Technion campus in Haifa and to establish the Technion Guangdong Institute of Technology, a joint venture between the Israeli

Technion Institute and Shantou University in southern Guangdong province, China.

This donation is the largest ever made to the Technion and one of the largest made to any Israeli academic institution. It demonstrates Li Ka-shing's acknowledgment of Israel's leadership and potential in innovation and academic research, and his interest in promoting this expertise to the Chinese people.

ChemChina

As mentioned earlier, China National Chemicals Corporation (ChemChina) made a landmark acquisition in October 2011 of a 60% stake in Makhteshim Agan, a leading Israeli manufacturer and distributor of crop protection products. In 2014, Makhteshim Agan changed its name to ADAMA Agricultural Solutions. *Adama* means earth in Hebrew.

Fosun

Fosun, China's largest privately owned conglomerate, has actively invested in Israel since 2013. Its first acquisition was Alma Lasers for US$240 million. Alma Lasers manufactures laser, light-based, radiofrequency and ultrasound products with an integrated product portfolio for esthetic and medical applications. The firm has leading-edge R&D capabilities in the medical and esthetic field. Fosun also expressed its willingness to open a technology incubator to promote Israeli technologies and identify additional business opportunities.

Through its private equity arm Fosun Capital, Fosun acquired in June 2015 a 52% controlling stake in The Phoenix Insurance Group from the Delek Group for approximately US$486 million, valuing the insurance firm at more than US$930 million. In the past, Fosun had shown interest in buying another leading Israeli insurance company, Clal Insurance Enterprises Holdings[47].

Fosun demonstrated its interest in strengthening its medical technology portfolio by buying Alma Lasers' competitor, Lumenis,

---

[47]   Source: *Globes.*

in February 2015. Lumenis develops and sells products for esthetic treatments and surgery based on lasers and light energy, mainly for the urology and ophthalmology markets. The company was eventually purchased by China's XIO Group in June 2015 for US$510 million. Fosun also led a private placement alongside GE Healthcare and Pontifax in a US$12 million investment in Check-Cap, bringing Fosun's shareholding to 13%. Check-Cap develops X-ray capsules for diagnosing intestinal cancer.

Yingke

Yingke, China's second-largest law firm with 2,000 attorneys and thirty-six offices worldwide, began operating in Israel in April 2013 when it acquired the Israeli boutique law firm of Eyal Khayat Zolt, Neiger & Co. (EKZN), which specialized in high-tech and venture capital. The establishment of a leading Chinese law firm in Israel demonstrates the increasing activity of Chinese companies in Israel.

Shanghai International Group and Sailing Capital

In July 2013, Shanghai International Group, one of China's largest state-owned financial holdings, invested in Israeli innovation company Mobileye through its private equity arm, Sailing Capital. Mobileye develops advanced driver assistance systems. In 2014, Mobileye raised US$1 billion by listing on the NASDAQ.

Xiaomi

In August 2013, leading Chinese electronics company Xiaomi invested US$11 million in Pebbles Interfaces, which develops motion gesture technology, alongside other investors including German-based Bosch Group's venture capital arm and United States-based SanDisk.

Bright Food

In March 2015, China's second-largest food and beverage company, Bright Food, purchased from Apax Partners and Mivtach Shamir a 77% controlling stake in Tnuva for an estimated price of US$1.47 billion. Tnuva is Israel's largest dairy company. The firm expects to bring the latest innovation and know-how in the field of dairy products and milk to China's Bright Food.

Plateau

Plateau, one of China's leading private equity funds, has also been active in Israel. CEO Arina Dong and senior executives from the firm value Israeli disruptive technology to enhance the development of Chinese corporations.

Plateau team at Western Wall in Jerusalem

Yuanda Enterprise

In September 2014, Yuanda Enterprise Group, a leading Chinese multinational involved notably in environmental engineering, acquired for US$20 million Israeli firm AutoAgronome, which specializes in smart irrigation and fertilization systems.

Baidu

In October 2014, Chinese firms Baidu, Ping-An, and Qihoo360 took stakes in Israeli investment fund Carmel Ventures. In December 2014, Baidu invested US$3 million in Israeli firm Pixellot, which provides unique high-quality and affordable alternatives to traditional video capture and production processes.

Guanxi Wuzhou Pharmaceutical

In December 2014, Guanxi Wuzhou Pharmaceutical, a leading manufacturer of cardiovascular and gynecology drugs, invested US$3 million in Hebrew University's technology transfer biotechnology arm, Integra.

Ping An Ventures

In December 2014, Ping An Ventures, a leading Chinese venture capital firm under the Ping An Insurance Group, led a US$27 million equity round to finance Israeli firm eToro, a leading social network for online trading.

Forwin Holding

In January 2015, Forwin Holding, a leading Chinese conglomerate with diversified activities in water, real estate, banking, and heavy industries, invested in Israeli disruptive technology firm StoreDot. This start-up has developed super-fast battery charging capabilities with applications in the mobile and electric vehicle markets.

Dan Catarivas, Director of Manufacturers' Association of Israel, with
Guo Min and Wei Liyang, owners of Shenzhen-based Angel Group

Alibaba

The same month, Chinese Internet giant Alibaba made its first
investment in an Israeli firm with Visualead. Visualead has developed
a QR code solution and operates in the Chinese market. The deal
attracted the attention of both international and Chinese media,
which closely monitor Alibaba owner Jack Ma's business activity. The
transaction also sent a strong signal to Chinese corporations to pay
attention to Israel's global leadership in disruptive technology and its
unique ecosystem offering numerous investment opportunities for
Chinese players.

JFC Shenzhen Business Entrepreneurs' Association

Leading business entrepreneurs Liu Na, Zhou Ying, and Yong Hong
Zhang from Shenzhen city are active in Israel. They have built a strong
dialog with local stakeholders and generate investment opportunities
in the Israeli innovation ecosystem.

JFC Group from China in front of the menorah
of the Knesset (Israeli parliament)

Research and Development Centers

Huawei

Huawei established its first R&D center in Israel in 2004 with fifteen staff and an initial investment of US$2 million. It appointed Solgood Communications as its exclusive representative in Israel.

Haier

Haier, one of China's largest consumer electronics and home appliances companies, opened an Israeli development center in 2008 to source innovation in the consumer electronics and appliance fields, most notably in white goods, smart televisions, and medical care products technologies.

Alibaba

The company is actively looking to open an R&D center to fill the needs of its corporate group.[48]

Investment Through Funds

Chinese corporations and financial institutions have invested actively in Israel through diverse platforms including incubators (ZTE, Ping An Insurance) and venture capital funds. Leading investors include Lenovo, China Everbright, Renren, and Tencent.

Infrastructure Activity

China Harbour Engineering Company – Ports

In June 2014, Pan Mediterranean, an affiliate of the Chinese government's China Harbour Engineering Company, won the tender organized by Israel Ports Development & Assets Company to build the south port in Ashdod for some US$800 million. The work includes the construction of a main 800-meter quay with a water depth of 17 meters, a working quay, 1,500 meters of secondary breakwater, and storage and operating areas, as well as the extension of the existing main breakwater by 600 meters.

This port is of strategic importance for both Israel and China. Israel is expected to enjoy greater competition between its ports (Ashdod, Haifa, and Eilat), which will likely reduce the frequency and scale of dockers' strikes and increase volumes handled, thereby decreasing the price of imported foods and other goods for the consumer. For China, this maritime port located in the Mediterranean Sea will provide a platform to supply the European, Middle Eastern, and North African markets.[49]

---

[48]   Source: Israel21c.

[49]   Source: *Globes*.

Hainan Airlines – Airlines

Privately owned Hainan Airlines, China's fourth-largest airline, announced in January 2015 that a new route from Beijing to Tel Aviv will operate three times a week from September 2015. This will compete with Israel's national carrier El Al, which according to the Israel Airports Authority, recorded 54,000 passengers (including Israelis) on the Tel Aviv–Beijing route in 2014.

Israeli Ministry of Tourism figures show 34,000 Chinese tourists visited Israel in 2013. This figure is expected to increase dramatically in the coming years given the tremendous interest on the part of Chinese citizens to travel to Israel both for business and tourism. The launch of a new commercial air route between Israel and China provides the logistical catalyst.

China Communication Construction Company – Railways

In July 2012, Israel's Minister of Transport Yisrael Katz and China's Minister of Transport Li Shenglin signed a historic agreement to build a high-speed railway to Eilat, Israel's southernmost city. The main project includes the construction of a cargo rail line that will link Israel's Mediterranean ports in Ashdod, Haifa, and Eilat. This strategically important project will bring Israel significant opportunities in trans-shipping by connecting the Red Sea to the Mediterranean. For China, the project provides the opportunity to strengthen its foothold in this strategic trade route, and to position China Communication Construction Company (CCCC) as the lead participant in a high-profile infrastructure venture.

Given the estimated US$12 billion cost of the project, and despite China Development Industrial Bank's proposal to finance part of it, progress has been slow in executing this gigantic infrastructure initiative. Nevertheless, government bodies in Israel and China have developed and nurtured advanced dialog that serves the countries' joint interests.

Singapore, South Korea, India, Japan and Taiwan Activity in Israel

Singapore

Singapore has been very active since its independence in cooperating with Israel in various fields, but predominantly in homeland security, venture capital (Vertex), fintech, and more recently biotechnology and medical devices. Singapore's activity and investments in Israel are mostly driven by government authorities and agencies, such as Government of Singapore Investment Corporation, Temasek Holdings, Infocomm Development Authority of Singapore, SPRING Singapore, and, to a lesser extent, by corporations such as Singapore Telecommunications (SingTel).

Government Authority

In 1997, the Singapore–Israel R&D Foundation (SIIRD) was established by the Singapore Economic Development Board and the Israel Office of the Chief Scientist to promote, facilitate and fund joint R&D industrial projects.

In November 2014, Temasek Holdings invested alongside India's Tata Group in Ramot's Momentum Fund. Ramot is Tel Aviv University's technology transfer company, which invests in technology developed by the university's researchers.

In February 2015, Infocomm Investments, part of the Development Authority of Singapore, launched a US$200 million fund dedicated to investments in Israel, particularly in cyber security and financial technology.

Cap Vista, the strategic investment arm of Singapore's Defense Science and Technology Agency, under the leadership of Chee Wei, is also active in the Israeli market.

Landmark private activities and investments in Israel are outlined below.

SingTel

In March 2012, SingTel acquired Israeli mobile advertising solutions provider Amobee for US$321 million in cash. In November 2012, SingTel through its SingTel Innov8 fund, invested in Everything. me, which develops innovative HTML5-based dynamic mobile application platforms.

In June 2014, SingTel's Israeli digital advertising unit, Amobee Technologies, acquired Israeli digital content intelligence and marketing solutions developer Kontera Ltd. for US$150 million.

Kusto

In May 2014, Kusto, a privately owned international industrial holding company with varied interests in construction, real estate, building materials and energy, acquired Israeli Tambour paints company for about US$125 million.

South Korea

In 2001, the Korea–Israel R&D Industrial Foundation (KORIL-RDF) was established by the Government of the Republic of Korea and the Israeli Office of the Chief Scientist to promote, facilitate and fund joint R&D industrial projects. This fund has financed numerous joint projects in both countries.

Samsung

Samsung is the most active South Korean firm in Israel with a dynamic portfolio of investments, two R&D centers specialized in telecommunications and semiconductors, and investments through corporate and third-party venture capital funds. Samsung established an initial footprint in Israel when it acquired TransChip for US$70 million in February 2007. The firm specializes in providing CMOS imaging solutions used to build microprocessors.

In July 2007, Samsung purchased Boxee for an estimated US$30 million. Boxee develops interfaces for home streamers. Its technology enables users to record and store content in the cloud and provides easy access to Internet video for various applications.

In 2008, Samsung invested in MCL Micro Components, which manufactures metal base advanced multilayer packages and substrates for the electronics industry. This was Samsung's first corporate venture capital investment in Israel.

Since then, Samsung Ventures under the leadership of Gonzalo Martinez de Azagra has built a strong portfolio of leading Israeli innovation firms. It has stakes in several firms, including the following companies.

Company	Areas of Activity
EarlySense	Develops hardware that monitors patients' vital signs without attaching any devices to the body
StoreDot	Develops technology to recharge batteries in minutes
RePlay	Creates three-dimensional 360° videos
Mantis Vision	Develops advanced three-dimensional-enabling technology and solutions
Ronds	Provides instant group video and chat functionalities
Imperium	Provides mobile security solutions

In 2013, Samsung boosted its presence in the innovation ecosystem by creating a US$100 million seed fund focused on Israel and the United States. The fund invests in innovation related to components and subsystems in televisions, mobile phones, computers, and digital devices.

LG

LG Israel Technology Center, operated by the LG Group, opened its doors in 1999. The center identifies and evaluates new Israeli technologies to incorporate into LG's family of products. The LG Israel Technology Center has partnered with Israeli academic institutions to develop joint R&D programs.

POSCO and Daewoo

In October 2010, POSCO joined Israel's government-sponsored R&D program. In June 2010, Daewoo International was selected as the EPC (engineering, procurement, construction) to build the OPC Rotem 440-megawatt natural gas power plant in southern Israel. In November 2012, Daewoo Shipbuilding & Marine Engineering, a subsidiary of Korean giant POSCO, was selected by Tamar's gas partner, Noble Energy, and the Delek Group to build the infrastructure of the gas platform.

SK Group

The SK Group, one of South Korea's leading conglomerates, acquired Camero, a developer of ultrawideband (UWB) based technology, in 2011. Camero's radar-based technology generates images beyond obstacles, such as walls.

Hyundai and Kia

Hyundai and Kia are the leading automotive brands in Israel, with an unrivaled 24% market share in the first quarter of 2015.

India

State Bank of India

In 2007, the State Bank of India, India's largest bank, opened a branch in Israel to focus on clients involved in the Israeli–Indian relationship, including in the diamond and high-tech sectors.

Tata Group

In April 2013, Tata Group invested US$5 million in Ramot's Momentum Fund. Ramot is Tel Aviv University's technology transfer company, which invests in technology developed by the university's researchers. The Tata Group has also indicated its willingness to invest in other local venture capital funds to enjoy the benefits of Israel's innovation ecosystem.

Infosys

In February 2015, IT giant Infosys purchased Panaya for US$230 million. Panaya provides automation technology for large-scale enterprise software.

Japan

Softbank

Softbank, one of Japan's leading telecommunications and Internet companies, has actively invested in Israel since the 2000s. Key investments include:

Company	Areas of Activity
RealM	Develops multidirectional video broadcasting systems
Camelot	Developer of security software for communication networks
CyberArk	Eliminates the need for an integrator for sensitive information
Insightix	Provides smart systems for IT visibility and network access control
Taykey	Develops social networks and platforms
Saguna Networks	Offers mobile edge computing enabling faster mobile broadband

Yaskawa

In 2008, Yaskawa bought RoboGroup, which specializes in motion control technologies, for US$8 million. In September 2013, it invested in Argo Medical Technology, a provider of medical devices to help individuals with lower limb disabilities to walk.

Sony

In May 2013, Sony invested US$10 million in Rainbow Medical, a medical devices R&D developer. In October 2014, Sony Pictures Television Networks acquired 50% of the Dori Media Group (diverse television channels) for an undisclosed amount.

Rakuten

In February 2014, Rakuten acquired Viber's instant messaging and communication application and its 300 million global users for US$900 million.

Toyota and Toshiba

Toyota has an R&D center in Israel focused on information and communication technologies. Toshiba has an innovation center formed on the back of the 2013 acquisition of OCZ Technology for US$35 million.

Mitsui

Mitsui Global Investment, Mitsui's investment firm, has an office in Israel. Its active investment portfolio includes the following companies.

Company	Areas of Activity
Autotalks	Provides very large scale integration solutions for vehicle-to-vehicle communication
Eyesight	Offers software technology for touch-free interaction with digital devices
Kaiima	Develops the next generation of seed and breeding technology using non-GMO
Kalutura	An open source online video platform
Kaminario	Manufactures flash SSD for SAN storage
Mo'Minis	A mobile game and distribution platform
Valens	Provides semiconductor and HD base technology

Japanese corporations and financial institutions also invest actively in Israel through diverse platforms, including incubators and venture capital funds. Leading investors in Israeli venture capital funds include Jafco, Nomura, Hitachi, NTT, Mitsubishi, Tokyo Ink, Seiko, CSK Venture Capital, Nippon Investment and Finance, and JAIC.

Taiwan

Winbond

In March 2005, Winbond, a leading producer of semiconductors, opened an R&D center in Israel.

Government

In January 2010, the Taiwanese Government announced its willingness to invest in local venture capital.

Epistar

In August 2010, Epistar a LED chip producer, invested in Israeli firm Oree, which manufactures flat casing for the LED market.

Several leading Taiwanese electronics groups have been active in Israel, including Macronix (the world's largest manufacturer of ROM memory), Taiwan Semiconductor Manufacturing Company (the world's largest dedicated independent semiconductor foundry), Teco (a conglomerate specialized in medium-voltage motors), and Tecom (the telecom arm of the Teco Group).

Taiwanese corporations and financial institutions have also invested in Israel through diverse platforms such as incubators and venture capital, including electronic giant Acer.

CHAPTER 9

Israeli Activity in Asia

Israel has been active commercially in Asian countries since its founding. Initially focused on defense activities, Israel is now a leading provider of numerous products and services in the areas of high-tech, telecommunications, biotechnology, pharmaceuticals, and agriculture. Virtually all Israeli companies sell to Asia, which is considered a core market for exports alongside North America.

Israeli Trade with Asian Countries

China

Over the past twenty years, trade between Israel and China has boomed, reaching an estimated US$9 billion in 2014, compared with US$50 million in 1992. Israel's exports to China amounted to US$2.8 billion in 2014[50]. China is Israel's largest trading partner in Asia and its second-largest in the world after the United States. Israel provides China with technology and innovation expertise, particularly in the fields of water, irrigation, desalination, agriculture, renewable energy, cyber security, Internet, telecoms, digital media, and healthcare, as well as in traditional trades like diamonds.

---

[50] All trade figures provided by Israeli government sources.

One of the leading diamond firms leveraging the Israel–
Hong Kong connection is Carnet (co-founded by Michelle
Ong, pictured) in Hong Kong with Dalumi in Israel

India

Bilateral economic development has flourished over the years, rising to at least US$5 billion in 2014 from US$200 million in 1992. Israeli exports to India amounted to US$2.8 billion in 2014. Today, India is Israel's second-largest Asian economic partner, and Israel is India's ninth-largest partner. Israel mainly exports diamonds and precious stones, defense equipment, chemical and mineral products, and high-tech equipment. India provides Israel with diamonds and precious stones, textiles, metals, plants, and vegetables.

Singapore

Israel and Singapore have a long history of close cooperation in the economic field. In 2014, the bilateral trade relationship amounted to about US$1.5 billion. Israeli exports to Singapore totaled in the region of US$729 million in 2014. Today, Israel exports mainly high-tech equipment to Singapore.

Japan

Israel–Japan bilateral trade represented about US$1.9 billion in 2014, with Israel exporting primarily machinery, electrical and medical equipment, chemicals, and diamonds totaling US$725 million. Israel imports automobiles, consumer electronics and machinery from Japan.

South Korea

Israel–South Korea trade reached US$2 billion in 2014. Israel's exports to South Korea amounted US$625 million in 2014, consisting mainly of electronics, telecom, biomedical, defense, and security equipment, as well as chemicals, diamonds, and precious stones.

Taiwan

Israel–Taiwan trade came to almost US$1.3 billion in 2014. Israeli exports to Taiwan amounted to US$481 million in 2014, consisting mainly of chemicals, semiconductors, diamonds, software, electronic, medical, and telecom equipment.

Israeli Corporate and Institutional Activity in Asia

To provide an understanding of the variety and depth of Israel's business activities and investments in Asian countries, we highlight below a number of Israeli companies with strong ties to Asia.

Asian Infrastructure Investment Bank (AIIB)

In April 2015, Israel joined the newly created Asian Infrastructure Investment Bank as one of its fifty-seven founding members. This new banking consortium led by China with its headquarters in Beijing aims to support investments in Asian and emerging markets. The bank competes directly with other leading institutions traditionally largely influenced by the United States, such as the World Bank, the International Monetary Fund, and the Asian Development Bank.

Teva

Teva Pharmaceutical Industries is the world's largest generic drug manufacturer and one of the top fifteen pharmaceutical groups globally with sales of approximately US$20 billion in 2014. Teva has been extremely active in Asia, particularly in Japan and Korea. The company signed two agreements in 2014 with Takeda Pharmaceutical, Japan's largest drug company. The first allowed Takeda to commercialize Teva's treatment for Parkinson's disease, Azilect. The second allowed it to market Copaxone, an innovative, breakthrough treatment to prevent the relapse of multiple sclerosis.

Teva acquired Taiyo Pharmaceutical Industry for US$934 million in 2011. Taiyo is Japan's third-largest generic drug manufacturer with a portfolio of more than 500 products and a strong presence in all major channels of the domestic pharmaceutical market. Teva's acquisition also gave it access to Taiyo's strong R&D team, local regulatory expertise, and state-of-the-art production facility. In the same year, Teva paid US$150 million to acquire Japan-based Kowa's 50% stake in a generics joint venture.

In 2012, Teva established a joint venture with South Korean Handok Pharmaceuticals. Teva manufactures and supplies a wide range of affordable and innovative medicines using its global resource chain. Handok controls the sales, marketing, and regulatory side of this estimated US$14 billion market.

Israel Aerospace Industries

Israel Aerospace Industries (IAI), is the largest aerospace corporation in Israel and a global leader in the development, manufacture, and enhancement of commercial and military aerospace technologies, systems, and products. It develops and manufactures business jet aircraft that form part of the US Gulfstream family of business-jet aircraft. IAI is active in several markets in Asia including India, Singapore, and South Korea.

In February 2015, IAI teamed up with India's Alpha Design Technologies to produce and market mini-unmanned aerial systems to accommodate the operational needs of Indian customers. Production takes place in India, and both IAI and Alpha handle marketing initiatives.

IAI opened a Cyber Early Warning R&D Center in Singapore in 2014. The center, modeled on IAI's Cyber Accessibility Center in Israel, employs Singaporean research scientists, computer analysts, and engineers.

In 2009, IAI subsidiary Elta Systems won a contract with South Korea, estimated at US$280 million, to supply radar systems to the Republic of Korea Army.

Israel Aerospace Industries develops and produces sophisticated and advanced systems. With its widespread international footprint, it has a leading position in Asia and China

Amdocs

Amdocs, which we have highlighted before, is a leading provider of software and associated services to communication, media and entertainment service providers in more than eighty countries. Amdocs is particularly active in Asia in Singapore, India, and Taiwan.

SingTel Group, a leading Asian telecommunications company, selected Amdocs to implement a business transformation project in its key markets of Singapore and Australia in December 2014.

In India, Amdoc's Mobile Financial Services Solution was selected in 2014, in association with Triotech's solution, to enable the State Bank of India to offer mobile financial services to India's unbanked or under-banked population.

Taiwan's FarEasTone Telecommunications, a leading provider of mobile, fixed-line, and broadband Internet services, selected Amdocs in 2014 to modernize its charging and billing systems with real-time capabilities. Amdocs also facilitated the launch of 4G LTE services to the local marketplace.

Elbit Systems

Elbit Systems is a defense electronics company with activities in aerospace, land and naval systems, command, control, communications, computers, UAV, advanced electro-optics, communication systems and radio fields. The company's business activity in Asia is currently focused mostly in the Philippines and South Korea. In 2014, Elbit Systems was awarded a US$20 million contract to supply armored personnel carriers to the Armed Forces of the Philippines. In South Korea, in 2013, Elbit established a joint venture called Sharp Elbit Systems Aerospace, jointly owned by Sharp Aviation K and Elbit. It offers maintenance, repair, manufacturing and R&D for advanced military avionics. Later in the same year, Elbit Systems won a follow-on contract to supply advanced helmet-mounted display systems to Korean Aerospace Industries, which supplies the Surion helicopter to the Republic of Korea Army.

Rafael

Rafael, a leading Israeli and global manufacturer of defense systems, specializes in underwater, naval, ground, air, and space systems. In February 2015, the company announced the establishment of a joint venture with the Kalyani Group of India to produce missile systems, remotely controlled weapons positions, and advanced systems for the protection of tanks.

Rafael acquired 49% of South Korean communication technology developer PineTelecom in 2013. Pine Telecom specializes in wireless data links for military and commercial applications.

Strauss

Strauss, one of the largest food and beverage firms in Israel, holds significant positions in international markets in the coffee and water segments. In China, in 2011, Strauss Water created a joint venture with Haier Group in consumer goods to tap the Chinese home-solution water market. Haier Strauss Water launched its first products in China under the WaterMaker brand.

Nice Systems

Nice Systems, Israel's global leading player in the field of telephone voice recording, data security, and surveillance, is active in several industries including finance, telecommunications, healthcare, retail, and utilities. In 2014, Nice was selected to deploy its Safe City solution in India's Nanded City, Pune to protect citizens, visitors, and historical sites.

In Singapore, Nice was awarded the 2013 Frost & Sullivan Asia Pacific Market Share Leadership Award for its workforce management systems. In 2012, Nice reached approximately 25% market share for these systems in Asia Pacific.

PT Bank Permata, a leading private bank in Indonesia, selected Nice in 2013 to install its service-to-sales, workforce optimization, and PCI compliance solutions. These flexible solutions help the bank meet changing customer and business demands while lowering costs.

In 2011, Nice was selected to install its IP video security solution within the subway networks of twenty-four major cities in China, including the Tianjin Metro. In 2009, it was selected to deploy IP video solutions and support content analytics for the Beijing metro lines. Nice also won China's Szechuan Province Chongqing Metro contract in the same year, providing digital video security solutions for eighteen monorail stations to protect passengers against crime and potential threats. In Malaysia, meanwhile, in 2012, Nice Systems deployed its remote banking fraud solution systems at CIMB Bank.

Netafim

Netafim is the world leader in smart drip and micro-irrigation. It is also a global player in agriculture, greenhouse turnkey projects, and biofuel energy crop markets. In Asia, Netafim has focused mainly on China, India, Japan, South Korea, and Thailand. The company plans to open a production line in 2015 for drip-irrigation equipment in China's Ningxia Hui Autonomous Region to better serve the needs of local farmers and government-owned companies in the field of agriculture.

Netafim has a well-established presence in India. In 2014, Netafim Irrigation India was named Best Irrigation Solutions Provider of the Year by Brands Academy, a premier brand management consultancy in India. The firm's subsidiary has been working on one of the world's largest drip-irrigation ventures: a 30,000 acre, twenty-two-village project in southern India involving more than 6,700 farmers.

Netafim, the pioneer and global leader in drip and micro-irrigation, maintains a strong position in Asia, including in India and China

IDE Technologies

IDE Technologies is extremely active in Asia, especially in China, where it built China's largest desalination plant in Tianjin, and India, where it built India's largest desalination plant for Reliance Industries.

IDE Technologies desalination plant in Tianjin, China

Orbotech

Orbotech is a leading producer of automated optical inspection and computer-aided manufacturing systems. This global company is particularly active in the Asian region, which is home to most of its largest electronics manufacturing facilities.

In China, in 2014, Orbotech was selected by Nanjing CEC Panda LCD Technology to provide its inspection, testing, and repair equipment solutions for its latest TFT LCD panel new generation fab.

In 2013, Orbotech signed a frame agreement with a leading Taiwanese-based printed circuit board manufacturer for the mass production of high-density interconnect printed circuit boards.

Mellanox

Mellanox is a global provider of Ethernet and InfiniBand computer networks and switches, and host bus adapters. The firm has significant activity in Asia.

In Japan, in 2014, PFU Limited, a Fujitsu company, selected Mellanox Open Ethernet switch systems for its high-definition video systems. In November 2014, Toshiba chose Mellanox 40 Gigabit Ethernet NIC for storage platforms. In September 2014, Mellanox Technology deployed its InfiniBand solutions for database acceleration at Yahoo! Japan at several of its data sites.

Matrix

Matrix, one of Israel's leading IT service companies, opened a new training center in 2014 through a joint venture between its John Bryce Hi-Tech College and the PTL Group. The training center is in Nanjing, China and operates in collaboration with the Nanjing Quality & Inspection Center. Matrix had already opened a mobile application development center in Changzhou, China the previous year.

Bank Leumi

Bank Leumi is a leading Israeli bank with activities in the United States, Switzerland, and the United Kingdom. In May 2014, the bank opened a representative office in Shanghai to serve as a financial bridge between Israel and China.

Kenon Holding /Quoros

Kenon Holding, a spin-off of Israel Corp, owned by Israeli business magnate Idan Ofer, has invested more than US$650 million to establish a joint venture with Chinese car manufacturer Chery Automobile. The venture will create and develop Quoros' range of sedan cars for the Chinese and European markets. Quoros sold about 7,000 cars in China in 2014.

Small Is Beautiful

A number of Israeli start-ups and smaller firms are active in various Asian markets. We highlight some of these companies below to illustrate the strategic importance of Asia in their business models.

Ceragon

Ceragon provides high-capacity microwave Ethernet and TDM wireless backhaul to wireless service providers and private companies. Ceragon is active in Asia and particularly in India. In 2013, the company won a contract from India's third mobile carrier, Idea Cellular, to modernize thousands of wireless backhaul connections around the country to support the data growth expected from 3G and future technologies.

Bio-Nexus

Bio-Nexus is a game-changing provider of IT workflow processes for various industries, including medical and airlines. The firm is very active in Asia with prestigious clients such as Eva Air, Qantas, and China Airlines.

Aero-Nexus' workflow platform

Traffilog

Traffilog is a leading Israeli vehicle management solutions developer. In 2013, it signed a landmark transaction with China's biggest bus company, Jiao Yun, which operates a fleet of 9,000 buses and 2,000 rescue vehicles. Traffilog's solution detects problems in vehicle engines and breaks, reduces maintenance costs, and improves efficiency.

D-Pharm

D-Pharm is a specialty pharmaceutical company that develops innovative drugs for the treatment of the most devastating brain disorders. In 2011, the firm awarded an exclusive development and commercialization license for its epilepsy treatment to one of China's largest drug companies, Jiangsu Nhwa Pharmaceutical Company.

Taditel

Israeli automotive electronics company Taditel, part of Ha'argaz Group, opened a plant in February 2015 in the CI3 industrial incubator in China's Changzhou Wujin Economic Zone. Taditel specializes in advanced voltage regulators used by leading global automotive makers including BMW, Audi, Volvo, Ford, GM, Toyota, Peugeot, Citroen, and Fiat.

Brainsway

Israeli medical device company Brainsway signed a distribution agreement in 2013 with Century Medical, part of the Itochu Group, to distribute Brainsway's Deep TMS system for treating major depressive disorders in Japan.

Tvinci

Start-up Tvinci signed a multimillion dollar transaction in 2013 to integrate its technology in the broadcasting of Singapore's MediaCorp.

Tvinci manages content broadcasting, protects content, and offers social television as well as a high-level user experience.

Pluristem

Israeli stem cell developer Pluristem Therapeutics formed a strategic partnership with South Korean CHA Bio & Diostech in 2013. The agreement encompasses the use of Pluristem's PLX=PAD cells in the treatment of two subsets of peripheral artery disease (PAD) in South Korea.

AB Dental

Israeli company AB Dental manufactures dental implants and computerized planning systems. In 2014, the firm signed a distribution agreement with Taiwan's BenQ Corporation. A joint venture named BenQ Dental Healthcare was established to distribute AB's dental implants among BenQ's dentists in the Asia-Pacific region.

Lexifone

Lexifone, which has developed a technology for simultaneous translation of telephone calls, signed an agreement in 2013 with the city of Changzhou in China to establish an R&D center. Lexifone's systems provide translation for speakers of different languages talking over the telephone or in face-to-face meetings.

IQP

Israeli start-up IQP has created a platform that allows users to make applications without any programming knowledge. It teamed up in March 2015 with Japanese corporations Fujitsu, NEC, and KDDI to develop applications for the Internet of Things (IoT). IQP's platform capabilities could dramatically change the way IoT operates in the automotive, energy, healthcare, smart home, smart city, cellular, and education fields.

CHAPTER 10

Three Relationship Pillars: Business, Tourism, Spirituality

Business, the Gate Opener

From the Silk Road to the Innovation Highway

Trade and business have always been gate openers for civilizations to come together, create understanding, and drive progress. From the ancient Silk Road to the new Innovation Highway, today's partnership in innovation and exchange between Israel and China has been nourished by connections spanning centuries, mutual respect, and shared interests. At the forefront of global research and development in many fields, Israel invents ideas and technologies that have been revolutionizing the world. China, which has no competition in the area of manufacturing, has already identified innovation as its Achilles' heel. Even though it has made tremendous progress in increasing the number of scientists and engineers it produces, China knows it needs more than this. Today's "match made in heaven" has been nurtured for centuries yet was not readily foreseeable only a few years ago.

The Tip of the Iceberg or, More Precisely, the Top of the Sand Dune

With investor interest and new business development at an all-time high and rapidly expanding, Israel and China's honeymoon is likely to extend far into the future. Today, China's most innovative technology giants, such as Alibaba and Baidu, are the trail-blazers and opinion leaders. While they establish their footprint in Israel, many more entrepreneurs and businesses are learning about Israel's innovation capabilities and preparing to harness its potential. These enterprises will soon join the early adopters of Israeli technology.

A similar trend is visible in other Asian countries. Major corporations from India, Japan, Singapore, and South Korea – including Infosys, Tata, Sony, Rakuten, Temasek, SingTel, Samsung, and LG – have pioneered investment in Israel's technology. They too are showing the way to others and helping to strengthen the bonds between Israel and Asia.

Tourism, Democratization of the Relationship

Fascination, Curiosity, and a Positive Image Motivate Chinese to Visit the Holy Land

Alongside the ongoing development of business links, the general Chinese population has demonstrated a growing interest in Israel's rich history and unique tourism opportunities.

China is fascinated by Israel from many points of view. First, from a historical perspective, China recognizes Israel as another of the oldest living civilizations that overcame great adversity to build a thriving country with a strong and powerful economy. While China and Israel are similar in this respect, their striking differences in size creates another angle of interest for the Chinese, who compare their 1.4 billion inhabitants and 9.6 million m^2 of land expanse with Israel's 8.3 million population and 0.021 million m^2 of territory.

Israel's prowess in the areas of academic, scientific and cultural advancement further fuels the interest. For example, the success of the Jewish people in winning the Nobel Prize intrigues the Chinese when they realize that while the Jews represent less than 0.25% of the world's total population they have been awarded more than 24% of the Nobel Prizes. No other nation has produced such outstanding results. In China's eyes, Israel appears synonymous with intelligence, success, and even miracles. The "early wave" of Chinese tourists in

Israel are thus motivated to understand the dynamics of this ancient nation, which is today at the forefront of global innovation.

The last key factor explaining the increasing influx of Chinese tourists to Israel is the pilgrim journey. The Chinese Academy of Social Sciences estimated the Chinese Christian population at approximately 50 million in 2012. These Christians provide an important source of present and future Chinese tourists wishing to connect with the Holy Land. While the pilgrimage to the Holy Land is often associated with Christians, it is not exclusive to any religion or culture. The Chinese are enthusiastic about visiting the Holy Land and its many UNESCO World Heritage Sites.

Tourism Is Blossoming

China's tourism is burgeoning. China became the world's largest source of tourism in 2012. More than 116 million Chinese ventured abroad in 2014, spending approximately US$120 billion. The Israeli Ministry of Tourism estimates that the figure will rise to 200 million by 2020.

Israel holds a strong attraction for tourists, who are drawn by its beaches, historical religious sites, UNESCO sites, culinary experiences, trendy nightlife, healthcare and high-tech expertise, business opportunities, and academia. Not surprisingly, these attributes have enticed an increasing number of Chinese tourists. Twenty thousand Chinese tourists visited Israel in 2012, increasing dramatically to 34,000 in 2014. The figure is forecast to reach 100,000 by 2017[51].

The Market Is Opening

Airlines, hotels, travel agents, and other industry participants view this market opening as a tremendous opportunity. They are actively preparing for a massive new influx of Chinese tourists in the coming years. It is not surprising to see Chinese-based Hainan Airlines opening up a Beijing–Tel Aviv route, initially with three flights a week, in September 2015. Hong Kong-based Cathay Pacific will follow with a

---

[51] Source: Israel Ambassador to China, Matan Vilnai.

Hong Kong–Tel Aviv route expected to open in mid-2016. Likewise, it is only a matter of time before cities such as Guangzhou and Shanghai provide direct access to Israel.

Spirituality, the Continuous Search for Wisdom

Common Values

Given the increasing exposure of the Chinese people to Jewish culture, as well as their respective ancestral traditions, one key theme beyond business or tourism has attracted Chinese interest in Israel. This is the quest for spirituality – the continuous process of learning and seeking out infinite wisdom.

While we have chosen not to address religion or faith, which we consider a private matter, we would like to emphasize some of the values and principles that the Chinese and Jewish civilizations share. These include the importance of the family unit, respect for the elderly, high regard for education, strong work ethic and integrity, the pursuit of happiness and success, and the importance of seeking wisdom and spirituality. Although the total size of the Jewish population is about 1% of the Chinese people, Jewish philosophy is rich and abundant, and surprisingly Jewish and Chinese philosophy share many points in common.

A personal experience of the authors serves to illustrate this. When we were in Israel recently with a group of Chinese investors, we took them to Masada, an ancient hilltop fortification symbolic of resistance to the Roman Empire. There is a religious ritual that Jewish men undertake on a more or less daily basis that involves putting on "tefillin", a set of phylacteries, one on the arm and one on the head. While we were engaged in this ritual, two of our guests pointed out that the specific areas covered by the tefillin were exact Chinese acupressure points.

Rabbi Adin Steinsaltz: Leading Authority on Jewish and Chinese Philosophy

Jerusalem-based Rabbi Adin Steinsaltz is one of the leading rabbis, scholars, and intellectuals in the contemporary Jewish world. Rabbi Steinsaltz has been hailed by *Time* magazine as "a genius whose extraordinary gifts as scholar, teacher, scientist, writer, mystic, and social critic have attracted disciples from all factions of Israel society."

Rabbi Steinsaltz also is the leading expert on comparing and contrasting Jewish and Chinese thought. He translated and published one of the most important books of the Jewish tradition, *Pirkei Avot* ("Ethics of the Fathers"), into Chinese. We share with you below one of Rabbi Steinsaltz's visionary lectures, which he gave in 1996 in Beijing, China, comparing Jewish and Chinese philosophy and culture. The message reveals that not only do Jews and Chinese have many values and principles in common, but these two ancestral people also have very successfully managed to preserve and apply their ancient wisdom and tradition in today's modern, innovative world.

Rabbi Adin Steinsaltz's "Ethics of the Fathers" book translated into Chinese, published by the Israel Institute for Talmudic Publications

Rabbi Adin Steinsaltz's Lecture from Beijing

The following are edited excerpts from a lecture given by Rabbi Steinsaltz in Beijing in 1996. This edited version has not been reviewed separately by Rabbi Steinsaltz.

"*Pirkei Avot*, or 'Ethics of the Fathers,' was composed in the Talmudic period. Talmud itself, in terms of quantity, is several thousand times larger than this book. Therefore, this book should be considered not as a summary of our culture, but merely a tiny sample of it. It is a book that is part of another, much larger, book, which itself is only a part of what was written in one of the most creative periods of our culture.

I chose to have this particular book translated into Chinese because it is, if I may say so, the most 'Chinese book' of our culture. One can find hundreds of parallels between *Pirkei Avot* and ancient Chinese culture – not only in general ideas but also in small details. Indeed, when I first encountered 'The Four Books' of Confucianism for the first time, many years ago, I filled up several notebooks with comparisons, sentence by sentence, between Confucius, Menzius, and, later on, also Lao-Tse, and the Jewish literature.

Sometimes the similarity is so striking that one might think that our respective Sages copied from each other. However, although these works were written at about the same time, the geographical distance between the two cultures was too big to make such copying possible. Indeed, only very few members of our people ever came to China in the past, and the tiny Jewish settlement that existed around Kaifeng for a few hundred years was very small, and neither very cultured nor influential.

The Chinese and Jewish peoples are similar in more than one way. We are both ancient peoples and, hopefully, also wise people. In addition, our cultures are non-missionary. We Jews do not send people to other countries to make them become Jews. We are just not interested in that. I also do not know about Chinese missionaries going to India or to any other country to convince, or force, the local people to become Chinese. Just as the Chinese believe that Chinese culture is basically the culture of the Han people and whoever else wishes to join, so we, the Jews, consider Judaism to belong to the Jewish people and whoever wants to get involved. In contrast, in the course of the Talmudic period of our culture, there sprang two outgrowths of Judaism – Christianity and Islam – which, being missionary religions, became far bigger in number than Judaism from which they originated.

Perhaps because of this tendency, both the Jews and the Chinese were accused by others of being too proud. But however this bent of ours may be termed, our sense of being self-sufficient, and our having no drive to spread our respective cultures aggressively, is also one of the basic reasons why we can have good dialogue. We may be interested in each other, intrigued by each other, we can surely learn from each other, but we are not trying to convert each other.

Rabbi Adin Steinsaltz with Philippe Metoudi, Lionel
Friedfeld, and Barry Topf (from right to left)

Generally speaking, Jewish and Chinese cultures have great admiration for wisdom. True, almost every culture claims to respect wisdom. However, both the Jewish and the Chinese cultures developed a form for this admiration, which is not found elsewhere. It may be called – a term that does not yet exist in any textbook of sociology – 'Sophocracy' or 'The Rule of the Wise'. Namely in both cultures, the Masters of Wisdom were not only scholars sitting in houses of study, learning and teaching, but they were also people who ruled society in a certain sense.

The Talmud describes a system whereby a committee of seventy Masters of Wisdom who sat in the capital, and who made all the principal decisions about law, order and behavior for the entire nation, basically ruled the country. In each region of the country, there were smaller committees of twenty-three Masters who ruled that region and smaller cities were ruled by committees of three. If an outstanding scholar were found in one of those committees of three, he would be promoted to that committee of twenty-three, and if a member of a committee of twenty-three showed outstanding ability, he would eventually be adopted by the highest committee of seventy. This is very similar to the Mandarin system that existed in China for many years.

In both the Chinese and Jewish cultures, the Wise Man is not famous merely for being creative, nor is he isolated from the society around him. First and foremost, the Master is supposed to be very well versed in classical literature. Interestingly, although Hebrew alphabet is one of the most ancient ones in the world – more than 3,000 years old – Jewish Sages had to remember thousands of books by heart.

All in all, the notion of wisdom, and the problems involved in it occupies an enormous part of *Pirkei Avot*. One of the central questions is, what is the relationship between theory and practice? What is the proper place for each of the two?

The Jewish answers to this question all go in one direction. We believe that theory that cannot be put into practice has no value. In *Pirkei Avot* chapter three, section 17, there is a series of paradoxes. It says: 'If there is no meal, there is no wisdom, and if there is no wisdom, there is no meal.' So where should one begin? Later on it says 'Anyone whose wisdom exceeds his good deeds, is like a tree that has big branches and short roots, so that when the wind comes, he is uprooted entirely, and a man whose good deeds exceed his wisdom, is like a tree that has deep roots and little branches, and this tree, even when a big wind comes, will always stay in place.'

Thus, a person, who does not practice what he says is not considered a scholar and is not worthy of any honor. We say that the Master should be the embodiment, the living example, of what he preaches. The Master is therefore constantly watched by his disciples, because he teaches not only when he lectures, but in each and every moment of his life, so much so that even a joke that the scholar makes is material for study. Indeed, the Master is compared to a tree in a wider sense as well, his fruits are his direct teachings, but like the branches, leaves and other parts of the tree, all his other aspects, too, are useful and beneficial.

One major difference between the *Pirkei Avot* and most of the authoritative books of Chinese culture is that in Jewish books, morality is an objective, eternal, transcendental, religious value, and not a matter of social convenience, whereas in Chinese culture it is basically a social and practical matter. And the question that I would like to pose here is, can society live on a practical morality alone? What should be the relationship between the Master of Wisdom and the powers to be?

As you know, Chinese culture is, in many ways, autonomous and, at least in ancient times, had little interchange with other cultures. The one famous exception is the Chinese attitude towards Buddhism. Chinese scholars went all the way from China to India, studied Sanskrit, and translated huge amounts of Buddhist texts to encompass a completely new culture and bring it to China. So in Buddhism, too (in its pure form, not in the popular blend of Buddhism, Confucianism and some Taoism that was created in China), there is a similar notion of objective, outside morality, which is not practical, of good and evil which are not social parameters, but they are objectives, eternal and binding, since they are not created by man. In the Jewish culture, however, it goes even farther than in Buddhism, as will be exemplified later.

As I said earlier, in both Chinese and Jewish cultures, the Masters of Wisdom were, on a certain level, the rulers of the country. But almost never, neither in our culture nor yours, did they have actual power. They were second to some other power. In other words, the Wise had wisdom, prestige, and great influence, but they did not have the army or the economy. In our history, there were figures who were masters both of wisdom and of the army, but they were the exceptions.

The issue of the proper relationship between the Wise and the Ruler, which has to do with traditional society in general and with the relationship between traditional and modern society, is discussed in 'The Four Books'. Confucius is essentially a supporter of the establishment, one may say – of almost any regime. Menzius raises the possibility of the sage saying 'no' to the ruler. Lao-Tse could, in modern times, be defined as an anarchist: he was not interested in the political system; he was working in ways that have nothing to do with it. Therefore, the relationship between him and Confucius, both historically and theoretically, is very interesting.

The Jewish scholars had an altogether different attitude to authority in general. They maintained that the king does not have supreme authority: he, too, is subject to the law, and if he disobeys, the people have the right to disagree with him, to disobey him, to revolt against him, even to kill him.

Finally, another question that I would like to raise is, what is, or should be, the relationship between the old and the new? Generally speaking, scholars tend to be conservative, both emotionally and culturally. Can, and should, the scholar act for new things and ideas, and work against the establishment?

In traditional societies, the scholar strives to be connected with the past, in which he believes, and with the status quo. As you know, a scholar is always happier to find an ancient text than a new one. In modern societies, it is almost the reverse situation, for the essence of modernity is the belief that the new is always better than the old. The greatest task of the modern scientist or researcher is to prove that what they write or do has never been said or done before. As academicians, you all know how much one sometimes has to work to prove that he did something new, even when there is nothing new about it.

This question is dealt with in *Pirkei Avot* as well. In Chapter two, section 9, there is a debate about which is the most praiseworthy scholar: the retentive one, who can be compared to a cistern of water that never loses a drop, or the creative one, who is likened to a spring that brings forth water. This is a modern problem as well.

All in all, it seems that in Jewish tradition, in general, there is something of a combination. On the one hand, everybody is very much urged to be innovative, but at the same time, the innovator is called upon to give the evidence that what his innovation does has already been done in olden times. Allow me to use a semi-jesting metaphor. In traditional Chinese philosophy, you have the ideal of an old woman dressed in an old dress. Modern culture opts for a young woman in a modern dress, which should be as short as possible. It seems that our tradition prefers young women in old dresses."

CHAPTER 11

Israel, China, and Asia: the Relationship in Fifty Years

Who would be presumptuous enough to say they could paint an accurate picture of the Israel–Asia relationship fifty years from now? In this region of the world, tomorrow is uncertain. For that reason, Israelis choose to focus their energy on the present. Indeed, fifty years seem like an eternity and almost no one, visionary or not, could hope to answer this question with accuracy.

So instead of reading a crystal ball, we have decided to take a practical approach. We highlight several trends that we believe will cement tomorrow's partnership. First we take the Israeli–Asian relationship drivers that are developing right before our eyes and project the main themes a few decades forward. With China as the driving force of the Asian region, we direct our analysis predominantly there.

The statements and hypotheses we derive from our approach are purely opinions and views and are not backed by any studies or surveys emanating from large institutions. In fact, the subject is still too new, too emerging to be filled by statistics and other forecasts. What follows is therefore our own "best guess".

Established Partnerships

In the business field, we expect the Israeli–Chinese relationship to mirror the intensity of the relationship between Israel and the United States in many ways, including the establishment of R&D centers for multinational firms, attraction to financial markets and financial institutions, infrastructure, and free trade.

Research & Development for Chinese Multinationals

Chinese multinationals have already begun to open R&D centers, or Innovation Labs, in Israel. ZTE, Huawei, and Haier are early adopters of this trend. Today, more than 250 multinationals have established R&D centers in Israel. About 66%[52], or 165, of these firms are American. In fifty years, given recent trends, we believe it is fair to say that the presence of Chinese conglomerates will at least equal the American presence. This means that hundreds of Chinese corporations will benefit directly from Israeli expertise in the next few decades. This figure will have tremendous significance both in terms of employment and cultural relationships between Israel and China. In fact, the whole Israeli ecosystem will be impacted by this phenomenon, starting with the service providers – such as lawyers, accountants, and auditors – that will play a role in ensuring a smooth adaptation to a new business environment.

Financial Markets

Israel and China have complementary value propositions. Israel is labeled as one of the most dynamic high-tech and start-up ecosystems in the world, but has failed to attract global liquidity to its financial market. China, on the other hand, with the world's largest savings ratio, is destined to manage tomorrow's global liquidity and needs a world-renowned high-tech label. The financial markets of Israel and China may be a perfect match.

China has gradually opened up. It continues to increase its presence in international financial market institutions, such as the International Monetary Fund (IMF), the World Bank, and the Asian Development Bank (ADB). When China created the Beijing-based Asian Infrastructure Investment Bank (AIIB) in late 2014, it sent a clear message to the world that it intends to be the leader in providing financing to Asian infrastructure projects. China indicated that the AIIB's activities would complement those of the IMF, World Bank, and ADB although some market participants regard it as a rival.

---

[52]   Source: *The Times of Israel.*

China's stated desire to make the renminbi –already the fifth-largest global currency in volume – an international and reserve currency will establish it as a world-leading currency, together with the US dollar and euro. The current equilibrium in the supply and demand of money will thus be impacted together with the whole financial system.

In fact, the rising power of China and its currency will reshuffle the cards in all market segments, including foreign exchange, equity, debt, and commodities, both for private and public market participants. The world of finance will change dramatically and Israel, which is very close to and dependent on United States equity markets – particularly the NASDAQ and NYSE – and venture capital for funding, will be heavily impacted.

This is only the beginning. More Chinese financial institutions and venture capital firms will enter the Israeli market in the coming years. In June 2015, China's Fosun Group bought The Phoenix Insurance Group from Delek Group[53]. It is only a matter of time before China's equity markets offer high-enough liquidity and valuations to convince Israeli entrepreneurs to list on China's stock exchanges, including the Hong Kong exchange, as an alternative to the United States.

We believe the Tel Aviv Stock Exchange has a unique opportunity to seize. It could privatize the exchange as other leading operators, such as ACE, EURONEXT, and NASDAQ, have done. It could also enter into a strategic alliance with a leading equity market operator, perhaps in Shanghai or Shenzhen, to create the new world technology equity platform of tomorrow.

Infrastructure

China is a critical and rapidly growing partner in Israel's transport and infrastructure sectors. China's role has evolved in a short time from a simple contractor to an operator and owner of strategic assets. Chinese companies such as Yutong Bus and Golden Dragon Bus are leading providers of buses and trucks in Israel. The Chinese

---

[53] Source: *Globes*.

government and a number of privately owned companies have played an active role in the construction of strategic Israeli infrastructure projects, such as the Carmel tunnels, Akko–Carmiel railway line, and future Ashdod port, as well as the operation of the new deep-sea port in Haifa.

In the railway field, Chinese firms will supply electric locomotives to Israel Railways and light rail cars to the Tel Aviv Metropolitan Area Mass-Transit System. China is likely to play an even larger role in the future flagship Eilat high-speed railway construction project, with Chinese contractors building the train line, Chinese manufacturers providing the locomotives and cars, and Chinese banks financing the cost.

China's "One Belt, One Road" strategy, announced by Chinese President Xi Jinping at the end of 2013, aims to create infrastructure links and commercial routes between China and the Middle East, Europe, and Africa via an overland belt and a maritime "road". Within this framework, China perceives Israel as an excellent trade hub. The future Eilat railway line will serve as a land bridge, linking the Red Sea from the port of Eilat to the Mediterranean Sea from the port of Ashdod, bypassing the Suez Canal. This new infrastructure will enable Israel to provide China with direct access to European, African, and Middle Eastern markets.

We expect Israel's and China's partnership in all areas of infrastructure – ports, roads, cargo, railways – to continue to advance over the next fifty years. This partnership will play a vital role in improving Israel's geopolitical stability and securing China's commercial trade routes between the Far East, Europe, and Africa.

Free Trade Agreement

Israel imported more from China than from the United States for the first time in its history in 2014. Data from the Israel Central Bureau of Statistics revealed that Israel imported US$8.1 billion in goods and services from China compared with US$7.4 billion from the United States. This likely marks the beginning of a long-term trend, which will be supported by the May 2015 signing of an agreement between Israel and China establishing an Authorized Economic Operator Program (AEO). The program enables fast-track customs clearance for the import and export of goods, thereby facilitating business trade between the two countries.

Furthermore, we believe that on a fifty-year horizon, Israel will have signed a free trade agreement with China given the importance of the strategic partnership in knowledge-based industry and industrial manufacturing. China state media has indicated that China will begin free trade negotiations with Israel in 2015.

Geopolitics

China recently has been such a driving force in Israel's economy that certain observers believe the country may be a perfect partner to offer viable solutions to the Middle East's troubled geopolitics. China's "One Belt, One Road" strategy, which aims at increasing trading and communication routes from China to Europe and elsewhere, could generate enough prosperity for the region's stakeholders to convince them to establish normal diplomatic relationships with Israel. China's growing economic and financial strengths, its excellent relationships and investment interests in various countries in the region, combined with its new appetite for international diplomacy, make it a credible force of influence for Israel and the entire Middle Eastern region in the coming fifty years.

Tourism

As leading China and Hong Kong-based airlines, hospitality, and tourism groups develop an interest in the Israeli market, many other players will follow. There is no doubt that Israel's unique tourism offering will attract an increasing number of Chinese tourists.

By 2017[54], more than 100,000 Chinese tourists are expected to come to Israel, and we foresee the number reaching as high as 500,000 visitors per year on a fifty-year horizon. In 2014, approximately

---

54   Source: Israel Ambassador to China, Matan Vilnai.

3.3 million tourists visited Israel. Of these, 626,000 hailed from the United States and 567,000 from Russia. Given China's large population of about 1.3 billion, we believe the country will easily be among the top three most active tourism groups in Israel.

To accommodate this new influx of tourism, we expect more foreign groups to enter the Israeli hospitality business with an aim to deliver first-class service. As more international hotel groups open five-star hotels in Israel, the services, food and beverage offerings, and prices will adjust to follow international trends. Israeli local competition will in turn renovate and build international hotels targeted to foreign clientele. Part of the food and beverage business will be tailored to Chinese tastes, from breakfast to lunch and dinner. Full-scale resorts with amenities such as spas, golf courses, and marinas will become a more prominent feature in this new hospitality landscape, targeting high-end Chinese and other international tourists as well as a growing local customer base.

Education, Sharing Values

Israel's top academic institutions, such as the Technion, Tel Aviv University, IDC, Hebrew University, Ben-Gurion University, and Bar-Ilan University, have attracted an increasing number of Chinese and Asian students in the past few years. Typical annual enrollments have risen from only a few students several years ago to several hundred students today. This figure is increasing at a double-digit rate. Non-profit organizations such as the Israel Asia Center support Asian foreign students in Israel with programs that empower them to become the next generation of Israel–Asia leaders.

Joint Academic Institutions

The Guangdong–Technion Israel Institute of Technology, a partnership between China's Shantou University and the Technion, is an excellent example of collaboration between the leading engineering school in Israel and one of the most respected academic institutions in southern China. Financed by the visionary Li Ka-shing, this project is the first large-scale bi-national education project carried out by Israel and China.

Other Israeli universities and Chinese academic institutions are working on bi-national projects. We expect many more universities to open partnering branches and learning centers in Israel and China in the years ahead. These will target bright and highly motivated Israeli, Chinese, and other international students with an aim to educate and train the next generation of leaders.

Partnership in Early Education

Both the Jewish and Chinese people place a high value on education. Accordingly, Israel and China have already adopted a common education system not only at university level but also at the high school, middle school, and early childhood levels. The Israeli-based Eastern Mediterranean International School proposes an Israel–Asia Youth Exchange in its 2015 curriculum.

China educators, schools, and private institutions view Israel as a fertile source of early education and youth programs. Many successful languages, arithmetic and logic interactive child solutions have been created by Israeli firms, such as UmaChaka Media's "TJ & Pals". These programs will be exported to China.

On a fifty-year timeframe, we believe that the partnership in education between Israel and China will expand dramatically, with bi-national summer camps, schools, universities and numerous continuous education programs and joint degrees awarded.

CONCLUSION

For the past few years, China has been investing heavily in Israel in various industries and market segments, from small innovative start-ups to large industrial companies.

Most of the economic commentators rightfully believe that China's rising economic power has enabled it to go shopping around the world to acquire the best products and know-how from various countries, and Israel is no exception.

We believe that the expanding strategic relationship between Israel and China, built up over the past few years, needs to be explained by factors such as history, culture, and spirituality. This way of looking at these recent phenomena is not so widespread in economic circles.

Israel and China are perfect partners in the new era of globalization. They share strong and complementary competitive advantages, with Israel contributing technology and innovation, and China supplying manufacturing and financial capability. Yet it is the people that make a relationship work, and these two countries have a tremendous respect for each other. This critical factor will help sustain the partnership.

The Chinese and Jewish people have known each other for thousands of years, lived together, and traded together. While it all began along the Silk Road, today the main business avenue is the Innovation Highway. Some things never change: Israel and China are still at the forefront of innovation and economic development. Common values and principles have brought these ancient nations together from ancestral times to the twenty-first century in a shared drive to respect tradition and innovate.

While building its new strategic relationship with Israel, China will likely be challenged by the United States' historically strong ties with the Jewish state, particularly in the business arena. Other Asian countries, such as India and to a certain extent technology savvy countries such as Japan, South Korea and Singapore also have ambitions to secure a preferred long-term partner status with Israel. There is no reason why all of the players on this stage cannot benefit from following the trends we have identified.

APPENDIX 1: MAPS

Silk Road Terrestrial and Maritime Routes

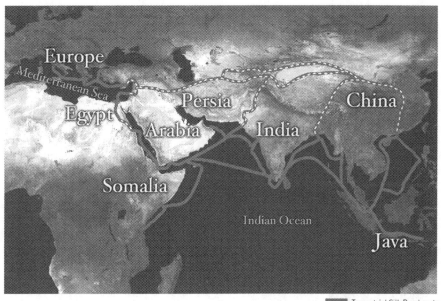

Terrestrial Silk Road route
Maritime Silk Road route

Kingdom of David and Solomon, Kingdom of Israel and Judah (1000-586 BCE)

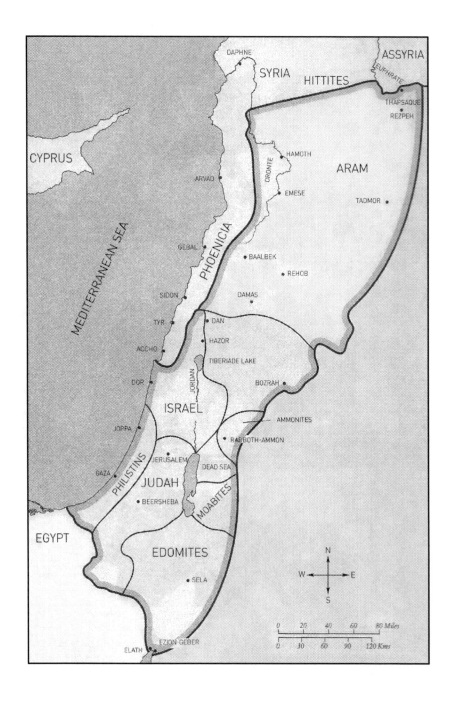

Migrations of the Northern Kingdom of Israel

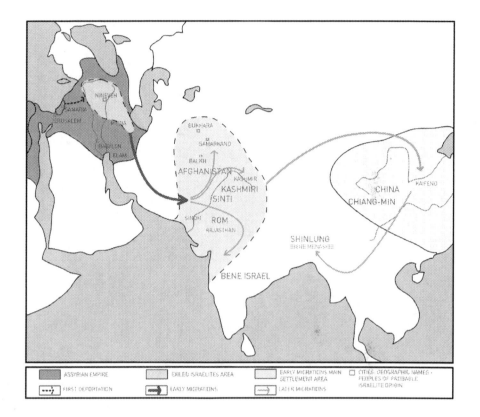

APPENDIX 2: ISRAEL'S HIGH-TECH SECTORS OF EXCELLENCE

Agricultural Technology

Perhaps more than any other sector, the State of Israel is associated in the popular imagination with pioneering innovations in agricultural technology (agritech). At its inception in 1948, Israel had almost no natural resources, with limited water supplies and little arable land due to the arid climate. Over time, Israelis developed techniques and technologies to increase crop yields and agricultural production. This pioneering spirit of innovation has carried over from generation to generation and has been a major contributor to Israel's leadership in agritech.

Examples of agritech innovators are highlighted in the table below.

Company	Sector	Areas of Activity
UniVerve	Energy	Develops fossil-fuel substitutes in form of microalgae oil
TransBioDiesel	Energy	Creates enzymatic process increasing profitability of biodiesel production from recycled animal fats and plant oils
SUBflex	Fishing	Designs submersible net cage system for offshore fish farming, enabling quality premium fish production
Morflora	Plants	Creates platform permitting modification of seed and plant genes

Company	Sector	Areas of Activity
Rosetta Green	Plants	Develops genetically modified potato and tobacco plants that can survive longer periods without water
Metabolic Robots	Poultry	Creates robotic food control kit for poultry sector, enabling control of frequency and quantity of meals by pre-set categories
CattleSense	Livestock	Provides non-invasive physiology sensor with solar energy and radio frequency capabilities, allowing critical information flow to cattle raisers
Kaiima	Crops	Develops new varieties of sustainable agriculture crops that improve productivity and land and water efficiencies

Clean Technology

Industry experts recognize Israel as a world leader in clean technology (cleantech). Israel became a cleantech powerhouse largely because most of its territory is arid and water resources were extremely limited. More than 500 Israeli companies specialize in solar and water technology. They have invested heavily in improving research and development in cleantech.

Israel's first Prime Minister, David Ben-Gurion, began to encourage the use of solar energy in 1955, when the government implemented the use of solar panels to heat boilers and provide hot water to individual homes. Many Israeli corporations also developed expertise in water technologies such as irrigation, desalination, filtering, and optimization management.

A number of firms have created breakthrough innovations in renewable energy and water – their accomplishments are shown in the table below.

Company	Areas of Activity
Renewable Energy	
BrightSource/ Luz	Pioneered solar thermal technology. Became Israeli R&D center of US-based BrightSource Energy
SolarEdge	Provider of power optimizers, solar inverters, and monitoring solutions photovoltaic arrays
Brenmiller Energy	Manufactures a modular solar steam generation product for power and steam applications
Tigo Energy	Optimizes the output of photovoltaic panels
Ormat Industries	World leader in converting geothermal and solar-heated steam into electricity
Arrobio	Waste energy processes eliminate need for prior separation of municipal solid water and have a recovery rate of 90%
Applied Cleantech	Recycles sewage to produce material used to manufacture ethanol
Phoebus Energy	Energy-efficient hybrid heat pump integrated into oil-based system, enabling 50%-70% oil saving and 80%-90% pollution reduction
PowerSines	Lighting energy controllers allow dynamic voltage regulation, power optimization, and 20%-35% electricity saving
Water	
Netafim	Pioneered drip technology – which saves water by allowing it to drip slowly into the soil through a system of pipes – for irrigation in the 1960s and now is the world leader in the industry

Company	Areas of Activity
Naan Irrigation Systems	Another specialist in drip irrigation technology; purchased by Indian firm Jain Irrigation, forming NaanDanJain Irrigation
IDE Technology	Specializes in water solutions, including development of desalination and industrial water treatment plants
Nirosoft	Utilizes advanced water treatment and desalination technologies for industrial and urban purposes
Desalitech	Specializes in manufacturing of seawater reverse osmosis systems
Amiad Water Systems	Highly regarded provider of filtration solutions
Bermad	Major manufacturer of valves and control management systems
Emefcy	Wastewater treatment firm producing low-cost electricity and hydrogen by using a bioelectro-chemical process with microbial fuel cell technology
Aqwise	World leader in development of innovative water and wastewater treatment solutions for municipal and industrial markets
TaKaDu	Provides integrated network management solutions, creating smart water grids that enable utility firms to preserve resources

Homeland Security

Since its creation in 1948, the State of Israel has had to defend itself from belligerent neighbors and terrorist attacks. Having fought major wars in 1956, 1967, 1973, and 1982, as well as numerous smaller clashes, it has had no choice but to develop a strong defense industry. This industry has innovated continuously over the years. From submarine, maritime, ground, air and aerospace to cyber technology, Israeli firms have become leading global players in the field of security.

The tables below highlight the primary activities of Israel's leading defense and cyber security firms and start-ups.

Company	Areas of Activity
Defense	
Israel Aerospace Industries	Leading aerospace and aviation manufacturer for both military and civilian markets. Developed first modern unmanned aerial vehicle (UAV) under leadership of Dr. David Harari
Rafael	manufactures high-tech defense systems for sea, land, air and space applications. It developed the Iron Dome defense system, which can intercept and destroy incoming rockets and artillery, under the leadership of Danny Gold, the head of the research and development bureau of the Israel Defense Forces
Elbit Systems	develops integrated battle systems for sea, land, air and space. It created the Long Range Reconnaissance and Observation System (LORROS), a sensor system providing long-range surveillance at all hours of the day and night
Aeronautics Defense Systems	Manufactures UAV systems
Magna BSP	Provides intrusion detection, recognition and tracking systems
Plasan	Leading manufacturer of vehicle armor systems

Company	Areas of Activity
Cyber Security	
Check Point	Pioneered the firewall market and specializes in IT security
Altor	Provides solutions for monitoring and enforcing security policies in virtualization environments; acquired by Juniper
Beyond Security	Provides tools to help manage security weaknesses in networks
Algosec	Provides software for network security policy
Anti-fraud	
Actimize	Provided financial crime prevention, compliance, and risk management solutions; acquired by Nice Systems
Cyota	Anti-fraud vendor; acquired by EMC's RSA
FraudSciences	Specialized in integrated systems for online transaction verification and fraud prevention; acquired by eBay subsidiary PayPal
ClearForest	Anti-fraud firm; acquired by Thomson Reuters
Intellinx	Provides solutions to record and analyze end-user behavior
BillGuard	Offers personal finance security service to analyze billing problems
Forter	Provides online merchants with solutions using behavioral analytics to identify fraudulent transactions in real time
AU10TIX	Provides document acquisition, authentication, and comprehension solutions to reduce identity fraud
Algorithmic Research	Offers digital signature solutions
Sentropi	Provides identification tracking solutions combining device fingerprint technology for online identification and fraud prevention

Company	Areas of Activity
Web security	
Trusteer	Provider of solutions for securing SaaS applications and sensitive web browsing transactions; acquired by IBM
CyberArk	An information security firm focused on privileged account security
Fireblade	Provides cloud-based security solutions for online businesses
Sentrix	Focuses on cloud-based website security solutions for various threats
Puresight	Gives parents tools to protect children against Internet threats
Imperva	Provides cyber and data security products
Commtouch	Provides technology solutions against spam
Data protection	
Varonis	Active in data governance
Adallom	Enables organizations to secure information in SaaS environments
Covertix	Permits confidential file sharing
Identity/access management	
SlickLogin	Provides authentication by smartphone; acquired by Google
Idesia Biometrics	Provides human identification solutions; acquired by Intel
Threat intelligence	
Aorato	Provides solutions protecting active directory services; acquired by Microsoft
Cyvera	Prevents targeted remote attacks on servers; acquired by Palo Alto Networks
Seculert	Provides cloud-based malware detection solutions
FortScale	Enables security teams to run big data analytics for cyber security

Company	Areas of Activity
CyActive	Forecasts future malware evolution based on bio-inspired algorithms
Morphisec	Provides multilayered security approaches to deter attacks on mission-critical systems
TopSpin	Identifies attacks in progress
Mobile security	
Discretix	Provides embedded solutions for device manufacturers to secure hardware, middleware, applications layers
Hermetic.io	Provides a mobile vault for securing digital assets like photos, bitcoins
Cellebrite	Provides mobile forensics solutions to law enforcement organizations
Physical security	
Magal	Provides physical and cyber security safety and site management solutions
Nice	Offers leading solutions for security intelligence, including surveillance, video recording, monitoring. Used for security at 2008 Olympics in Beijing
Verint	Offers leading solutions for security intelligence
Briefcam	Provides image processing technology that summarizes video footage. Used in Boston Marathon bombing to identify the terrorists.
Visionic	Electronic security systems provider; acquired by Tyco International
Camero	Developer of ultrawideband-based technology that generates images beyond obstacles such as walls; acquired by SK Group

Digital Media

The Israeli digital media industry has been very active since the 1990s, notably with the acquisition of ICQ (instant messaging) by AOL, but also with the numerous acquisitions of Israeli start-ups by American technology giants.

A listing of companies involved in various sectors of digital media appears in the table below.

Company	Areas of Activity
Applications	
Waze	World's largest community-based traffic and navigation application; acquired by Google
Onavo	Mobile data optimization; acquired by Facebook
Viber	Phone and messaging application; acquired by Rakuten
Dragon Play	*Live Hold'em Pro, Wild Bingo, Farm Slots*; acquired by Bally
Diwip	*Best Power, Best Blackjack*; sold to Imperus
KitLocate	Application enabling customers to add location features to Android / iOS; purchased by Yandex
Moovit	Community-based public transportation application
GetTaxi	Taxi ordering application
Pango	Mobile payment application for parking spaces
Slidly	Application enabling users to share photos and videos along with their favorite music
Magistro	Streamlines video editing
Glide	Instant video messenger
365Scores	Enables users to create sports channels
Drippler	Offers personalized news services
Seeking Alpha	Provides financial analysis and stock market news
eToro	Provides social trading services

Company	Areas of Activity
Perion Networks	Offers monetization solutions for mobile applications and desktop
Crossrider	Provides equivalent solutions with big data features
Appwiz	Offers developers a set of mobile application monetization tools
E-commerce	
Shopping.com	Leading price comparison service for online shopping; acquired by eBay
Fiverr	Provides marketplace for freelancers to offer services
EatWith	Offers platform connecting potential hosts with people who want to eat out; often referred to as Airbnb for food
Nipendo	Offers cloud-based platform enhancing buyer-supplier collaboration
Tag'by	Developed unique solution that enriches retailers' point of sale with dynamic social media platform
MySupermarket	Enables buyers to compare prices and shop online
Upstream Commerce	Provides retail pricing intelligence
FeeX	Helps users find and reduce fees for diverse financial accounts
SundaySky, Idomoo	Power personalized video experiences
Digital games	
Plarium	Largest player, with activity in social, mobile and browser-based gaming
SideKick	Develops titles for various game platforms
JoyTunes	Develops educational musical games programs
TabTale	Develops mobile games targeted at young children
888 Holdings, Playtech	Online gambling

Company	Areas of Activity
Digital advertising	
Matomy Media Group	World leader in digital performance-based advertising
XLMedia	Global digital publisher and marketing company
Marimedia	Leader in digital advertising monetization
DoubleVerify	Ensures that ads do not run on illegal content
MyThings	Offers large commercial brands ability to follow users on Internet and promote relevant deals
Kenshoo	Offers platform to improve search engine marketing results
Taboola, Outbrain	Dominate content recommendation market
TicTacTi	Develops solutions that enable publishers and game developers to increase revenues by displaying in-game ads in specific content
Double Fusion	Offers possibility to embed advertising in PC and video game consoles
TodaCell	Mobile advertising network catering to mobile advertisers and publishers
ClicksMob	Acts as affiliate network mediating between developers and traffic providers
Ubimo	Develops mobile advertising platform delivering location data to provide customers with advanced features
YouAppi, Appnext	Enables content providers to offer applications for download based on behavior relevancy

Semiconductors and Electronic Components

Israel has been at the forefront of the semiconductor industry and a world leader in fabless firms. Numerous multinationals, attracted by the high levels of Israeli expertise, have established R&D centers in Israel. These include Intel, Freescale, Marvel, and Texas Instruments. Intel developed many of its chipsets in Israel, including Pentium processors, Centrino, Core 2 Duo series, Sandy Bridge, and Ivy Bridge. Key Israeli companies include Mellanox, a provider of connectivity solutions for servers and storage, and Orbotech, a world leader in inspection and imaging systems for the electronics industry.

The table below highlights activity in the sector.

Company	Areas of Activity
Semicondutors / Electronic components	
Anobit	Flash memory chip designer; acquired by Apple
Primesense	Developer of 3D motion-sensing devices; acquired by Apple
Mellanox	Provider of connectivity solutions for servers and storage
Orbotech	World leader in inspection and imaging systems for electronics industry
Multimedia/entertainment	
Zoran	Leading developer of digital signal processing for entertainment and consumer electronics market
SURF Communication	Manufactures hi-density multimedia DSP processing boards
EyeSight	Offers gesture-recognition technologies that enable users to control various devices
Extreme Reality	Supplies consumer electronics OEM with integrated remote touch-free software interface
ZRRO	Provider of multi-finger near field 3D positioning technology for touch screens
Corephotonics	Develops dual-camera technology to enhance optical zoom
Advasense	Provides CMOS image sensor solutions for camera-phone market
Waves Audio	Provides audio processor plug-ins
Communication chips	
Altair	Developer of single-mode LTE chipsets

Company	Areas of Activity
Asocs	Provides modem-processing units that enable implementation of large variety of communication standards
Precello	Develops low-cost digital baseband processors; acquired by Broadcom
Wilocity	Offers one of fastest WiFi chips enabling data transfer of up to 7 Gbps; acquired by Qualcomm
Celeno	Develops components and subsystems for high-performance carrier-class WiFi systems
Others	
Nova Measuring Instruments	Specialized in metrology
Ophir Optronics	Specialized in instrumentation; acquired by Newport
Annapurna Labs	Offers ARM-based communication controllers
SolChips	Integrates low-power electronic devices to solar photovoltaic systems

Telecommunications

The Israeli high-tech industry has always been at the forefront of the communication and telecommunication sectors. It pioneered many breakthrough innovations, that characterize the sector today, such as VoIP, WiMAX, and TDMoIP.

Telecoms multinationals have been active in Israel with R&D centers and multiple acquisitions of Israeli companies. Cisco has been the most active, acquiring twelve Israeli firms for US$6.5 billion. Other multinationals with a strong presence in Israel are Motorola, Alcatel-Lucent, Broadcom, Qualcomm, Avaya, and Samsung. Israeli domestic incumbent communication groups include RAD and ECI Telecom. The remainder of the Israeli communication industry has been developing rapidly with many leading players.

The table below provides a flavor of what has been achieved in the sector.

Company	Areas of Activity
VoIP	
Vocaltec	Introduced first commercial phone software offering PC-to-PC communication
Deltathree	Global provider of VoIP telephony services, products, and solutions
Jajah	Provider of IP-based managed services; acquired by Telefonica
AudioCodes	Packet-based solutions for voice networks
Spikko	Provides free VoIP calls
EIM Telecom	International VoIP wholesale provider
Call Me	VoIP to fixed-line specialist
Video conferencing	
Radvision	Provider of telepresence and video-conference technologies; acquired by Avaya in 2012
Vidyo	Multipoint video communications software
Ethernet	
Actelis Networks	Ethernet over copper solutions
Telrad	LTE products designed to enable wireless broadband connectivity
Fibrolan	Integrated access systems and solutions for carriers, service providers and mobile operators
ECI Telecom	Products and solutions for the carrier Ethernet and IP networking
Orckit-Corrigent	Global supplier of telecommunications networking equipment
RAD	Carrier-class Ethernet access solutions

Company	Areas of Activity
Optical networks	
ECI Telecom	Provides full range of networking solutions to telecom providers
MRV	Provider of optical transport solutions
FiberZone Networks	Develops intelligent solutions for fiber-optic and optical network infrastructure
Packetlight Networks	Supplies multi-service optical transport and access systems
Video delivery	
Giraffic	Provides video acceleration cloud services to online video providers
QWilt	Offers network operators increased capacity solutions to improve subscriber experience
Contextrem	Innovates cloud-based IP services for video / multi-play telecom
Tvinci	Supplies cross-platform solutions that enhance premium content consumption on digital devices
Applicaster	Provides white-label broadcast solutions for cross-screen TV experience
Imagine Communication	Provides media software and video infrastructure solutions
LiveU	Offers portable video transmission solutions with live broadcast using multiple cellular connections and other data networks
Network/traffic management	
Allot Communication	Provides traffic management solutions
Ortiva Wireless	Provides mobile carriers with dedicated video optimization gateway to reduce wasted bandwidth
Oversi Networks	Provides rich-media caching and content delivery solutions for Internet video peer-to-peer

Company	Areas of Activity
Radware	Provides solutions to manage IP services
Intucell	Cellular optimization network; purchased by CISCO
Ceragon, Radwin	World leaders in wireless backhaul – the process of getting information from end-users to network distribution points
Celtro	Provides mobile-centered backhaul optimization
Wintegra	Developed software-based solutions for the mobile backhaul infrastructure market; acquired by PMC-Sierra

Information Technology (IT) and Software

Leading Israeli IT services firms include Matrix, Malam Team, Ness, and Taldor. These companies operate in different markets segments and specialize in providing full IT solutions services, including outsourcing, software development, cloud and cyber security consulting. They are active globally in Europe, the Americas, and Asia.

Multinationals have a strong presence in the IT infrastructure market segment owing to many acquisitions of Israeli firms. Major acquisitions include HP's 2006 purchase of Mercury Interactive in the IT management field for US$4.5 billion; IBM's acquisition of storage companies Storwize, XIV, and Diligent; EMC's purchase of Kashya and Illuminator in data protection, and nLayers in network management; and the landmark 2012 acquisition of XtremIO in the field of storage systems for US$430 million. VWware meanwhile acquired Digital Fuel, which provides IT cost optimization solutions, for US$120 million in 2011.

Innovative Israeli IT companies are highlighted in the table below.

Company	Areas of Activity
IT Services	
Matrix, Malam Team, Ness, Taldor	Specialize in providing full IT solutions services, including outsourcing, software development, cloud and cyber security consulting
IT Infrastructure	
DensBits Technologies	Provides IP and controller technology for NAND flash-based storage systems
Reduxio	Offers hybrid storage systems
Elastifile	Provides converged storage solutions for virtualized datacenters and private clouds
ScaleIO	Specialist in software solutions that enhance storage capacity in servers; acquired by EMC
CTERA Networks	Provides cloud storage gateways
Zerto	Develops disaster recovery and business continuity software for data centers and cloud environments
Precise Software	Specialist in application performance management; acquired by Veritas (now Symantec)
Uppspace	Provides anti-crash and performance monitoring solutions
Digital Fuel, Fixico	Provide SaaS-based IT management
dbMaestro	Provides database change management solutions
ScaleBase	Addresses database availability and scalability

When it comes to enterprise applications, a number of multinationals have reinforced their expertise by acquiring Israeli firms. Microsoft purchased WebAppoint (online scheduling), YaData (data mining), and Gteko (support automation). Oracle acquired HyperRoll (business intelligence). SAP meanwhile acquired TopTier from well-known entrepreneur Shai Agassi in 2001 for US$440 million.

Israeli innovators in software can be found in the table below.

Company	Areas of Activity
Software	
Amdocs	Operational support solutions in various areas, including billing, customer service, sales, and marketing
Bio Nexus	Mobile workflow processing software for real-time quality control
Captiza	Cloud-based solutions for migrating existing business applications to mobile platforms
ClickTale	Customer experience analytics solutions
CloudShare	Cloud platform for testing IT applications
Comverse	World leader in billing market for telecom players
CVidya	Revenue assurance and fraud risk management
EXAI	Website creation from Facebook pages
Experitest	Mobile testing solutions
Magic Software	Cloud and application platform solutions
Mercury Interactive	Performance application solutions
Nice	CRM
Perfecto Mobile	Solutions to access mobile devices through Internet
Pontis	Personalized and contextual marketing across various platforms and channels
Radview	Performance application solutions
Totango	Big data analytics and segmentation
Valooto	Cloud-based collaborative sales engagement platforms
Verint	CRM

Company	Areas of Activity
Webydo	Online website design software
White Source	Cloud-based open-source lifecycle management
Wix	Do-it-yourself websites and publishing solutions
Worklight	Cross-platform deployment for digital devices
Zend Technologies	Pioneered PHP language

Life Sciences

Israel's life sciences industry has seen rapid growth over the past decade, and it now plays an important role in world healthcare. Its commercial success has hinged on strength in academic research, as well as government support and an increased availability of funding.

Israeli firms at the forefront of the medical devices, digital healthcare, and biotechnology (biotech) sectors are highlighted below.

Company	Areas of Activity
Medical devices and Digital healthcare	
SHL Telemedicine	Offers personal telemedicine systems focused on cardiovascular-related diseases
Aerotel	Provides telemedicine solutions
LifeWatch	Originator of technologies to help physicians detect and analyze patient syndromes
dbMotion	Provides health interoperability solutions for connected healthcare; acquired by Allscripts
iMDsoft	Offers graphical clinical patient information systems for hospitals
FDNA	Platform facilitates detection of facial dismorphic features and recognizes patterns of human malformation from facial photos
Night-Sense	Provides non-invasive medical devices that enable real-time alerts to sleep hypoglycemia for diabetics

Company	Areas of Activity
Mediscope	Provides mobile digital companion for serious illness support
Sweetch	Offers platform to monitor and help prevent diabetes
Nutrino, MakeMyPlate, Fooducate	Applications promoting nutritional awareness, diet and wellness
Biogaming, Physihome	Active in detection and treatment of physical and neurological diseases
Genoox, Genome Compiler	Active in field of genomics
Medical Opinion, Second Opinion	Platforms focusing on patient-physician communications
Essence, Medilogi	Offer senior population monitoring
Medcon	Provider of cardiac and information management solutions; acquired by McKesson
Starlims	Information technology suite developer for laboratories; acquired by Abbott
Biotechnology	
Teva	Copaxone treatment for multiple sclerosis and central nervous systems disorders has resulted in 80% reduction in relapsing-remitting multiple sclerosis. Azilect Rasagiline-based product aids patients with Parkinson's disease by blocking breakdown of dopamine in brain
Exelon	Postpones worsening of Alzheimer's disease for up to one year for 50% of patients; drug developed by Novartis

Company	Areas of Activity
Regentis Biomaterials	Developed an innovative, injectable biosynthetic gel used to stimulate bone repair
TheraVitae	Developed a revolutionary treatment for heart disease that rebuilds heart tissue using patient's stem cells
Medinol	World manufacturer of innovative stent designs for heart catheterization
Compugen	Leading drug discovery firm creating protein and antibody therapeutics
Rosetta Genomics	Discovered micro nucleic acid RNA

Other Fields of Expertise

Printing

The global printing industry is well represented in Israel with leading companies outlined below.

Company	Areas of Activity
Scitex	Large digital industrial presses; acquired by HP
Indigo	Digital offset printing; acquired by HP
Nur	Wide-format digital inkjet printers; acquired by HP
Press-sense	Leading developer of workflow and management solutions for print providers; purchased by Bitstream
Objet	Global leader in 3D printing; merged with US-based Stratesys
Landa Corporation	Founded by Indigo's founder Benny Landa. Unveiled new digital printing technology based on nano-pigment that enables printing on almost all materials

Company	Areas of Activity
Scodix	Provider of digital print enhancement presses for graphic art industry
DigiFlex	Offers diverse inkjet-based solutions
Pzartech	Offers an innovative 3D design marketplace

Financial Technology

Israel's expertise in financial technology (fintech) has attracted the attention of major international banks, such as Citi and Barclays, which have opened Israel-based innovation labs. The table below highlights the activities of Israel's leading fintech players and start-ups.

Company	Areas of Activity
Fintech	
Sapiens	World leader in software development for insurance industry
Fundtech	Provides cash management and payment settlement solutions for financial institutions; acquired by US-based GTCR
Traiana	Offers over-the-counter foreign exchange, derivatives, and cash equities platforms; acquired by ICAP broker
Super-Derivatives	Produces multi-asset derivatives pricing and risk management tools; acquired by US-based ICE Intercontinental Exchange
Digital payments	
Credorax	Provides digital payment processing for online merchants
Zooz	Its platform shortens payment process
MyCheck	Provides mobile payment platforms
Payoneer	Streamlines corporate payouts by providing prepaid debit cards

Company	Areas of Activity
LogicalForm	Provides bitcoin-clearing systems for financial institutions
Hermetic.io	Develops security solutions for digital assets such as pictures or bitcoins on mobile phones
Trading tools / applications	
Final Israel	Provides traders with comprehensive software tools to monitor key financial parameters like bid/ask, volume
FMR Computers and Software	Provides front, middle and back office systems for stock exchange members
TradAir	Offers front-office optimization solutions and algorithmic trading
Strategy Runner	Provides server-based technology for algorithmic trading solutions; acquired by MF Global Holding
Foreign exchange	
TraderTools	Provides trading platforms for financial institutions with liquidity aggregation and price engine management tools
ForexManage	Supplies portfolio risk management and online trading platforms
Surecomp	Offers trade finance and treasury solutions and reconciliation with SWIFT confirmations

Automotive

A country in love with cars and traffic, Israel has become a leading center for technology in the global automotive industry. Israeli firms are active in different market segments, including safety with Mobileye, a provider of advanced camera-based driver assistance systems. Mobileye has raised more than US$500 million in private funding (of which US$400 million came from Goldman Sachs and Morgan Stanley), and more than US$1 billion in 2014 through its initial public offering on the New York Stock Exchange. Mobileye's systems will be integrated into Honda, BMW, Ford, Chrysler, and General Motors cars.

The table below showcases innovators in the industry.

Company	Areas of Activity
Mobileye	Provider of advanced camera-based driver assistance systems
Brightway Vision	Develops night-vision technology to make night-time driving easier
iOnRoad	utilizes smartphone cameras and sensors to detect vehicles in front of the car and alert in case of danger
I4drives	Enhances driving experience and safety by employing smartphone capabilities
MobiWize	Provides vehicle connectivity to reduce maintenance costs and enhance driving experience
Autotalks	Fabless semiconductor solutions that enable the exchange of information between vehicles and infrastructure
Arilou	Provides security solutions that block prohibited messages to vehicle's controller network
Pointer Telocation, Ituran	Offer location-based solutions for stolen vehicle tracking and recovery
E-Drive Technology	Provides fleet management and telematics solutions
Waze	World's largest community-based firm to which drivers contribute their first-hand experience; acquired by Google

Energy Management

Israeli firms are leaders in the energy management industry, especially where consumer energy management is concerned. Top firms in the sector are shown below.

Company	Areas of Activity
StoreDot	Has developed a revolutionary new generation of batteries based on nanodots, which enable a phone battery to recharge in one minute and are expected to charge an electric car in three minutes
Powermat	Provides wireless charging boards
Greenlet Technologies	Helps utilities control power consumption by giving incentives to end-users
Grd4C	Maximizes efficiency of energy operations by analyzing smart data
PowerCom	Provides smart grid solutions for electricity, water, and gas utilities
Lightaap Technologies	Develops solutions for industrial energy management
Meteo-Logic	Provides power and weather forecasts based on big data

Smart City

Israel is also a world leader in the smart city sector with innovation driven by the companies listed below.

Company	Areas of Activity
LeanCiti	Provides data analytics and visualization solutions enabling cities and citizens to make real-time decisions about resource usage
MobilityInsight	Optimizes smart city real-time transport management

Company	Areas of Activity
BreezoMeter	Location-based application displays air pollution levels in real time
Nisko Telematics Systems	Specializes in automated meter infrastructure systems for water authorities
Arad Metering Technologies	Provides automated wireless reading systems

Big Data

Israel firms are extremely active in the big data field with companies such as Panorama Software, a provider of SaaS-based analytics, reporting, and visualization designed tools. Alooma offers a platform with an easy data transformation interface, and C-B4 provides data compression and machine learning algorithms to enable automatic generation of prediction models.

Product Life Cycle Management

Israel has been active in the sector. SmarTeam, which provides collaborative product data management solutions, was acquired in 1999 by market leader Dassault Systems. VisualTao, a web-based computer-aided design (CAD) firm, was acquired in 2009 by Autodesk. In 2005, USG (now part of Siemens) purchased Tecnomatix, which develops digital manufacturing process planning and optimization tools.

BIBLIOGRAPHY

Sources about Israel, China and Asia are numerous. We have included herein the main articles, publications, and books and that were used as reference materials.

Newspapers, Magazines and Industry Reports

Bank of Israel
The Diplomat
The Economist
Financial Times
Forbes
Globes
Haaretz
Israeli Government and Ministries
Jerusalem Post
KPMG/IVC Research Center
The Marker
Moody's
S&P
The Times of Israel
Ynet news

Publications

Bufman, Dr. Gil Michael, Raz, Eyal, & Hager, Noach, The Potential of Natural Gas in the Israeli Economy, Bank Leumi (2014)
Gil, Naama & Tepper, Raz, Underlying Legal and Regulatory Framework of Office of the Chief Scientist of the Israeli Ministry of the Economy, FCB Lawyers (Jan 2014)

Katz, Nathan & Goldberg, Ellen S., The last Jews in India and Burma, Jerusalem Center for Public Affairs, Jerusalem Letter (1988)

Raviv, Oren & Yachin, Dan, ICT Industry Review, Israel Advanced Technology Industries (2015)

Trajtenberg, Manuel, R&D Policy in Israel: An overview and Reassessment, Tel Aviv University (2000)

Reference Books

Abadi, Yakob, Israel's Quest for Recognition and Acceptance in Asia, Garrison State Diplomacy, Routledge (2004) ISBN 0-7146-5576-7

Corfield, Justin & Corfield, Robin, Encyclopedia of Singapore, Scarecrow Press: (2006), ISBN 0-8108-5347-7

Dommen, Arthur J., The Indochinese Experience of the French and the Americans: Nationalism and Communism in Cambodia, Laos, and Vietnam, Indiana University Press, (2001) ISBN 0-253-33854-9

Edelsheim, Alfred, History of the Jewish Nation after the destruction of Jerusalem under Titus, Kessinger Publishing, (2004) ISBN 1-4179-1234-0

Ehrlich, Mark Avrum (ed.), Encyclopedia of the Jewish Diaspora: Origins, experiences, cultures, volume 1, ABC-CLIO (2009), ISBN 9781851098736

Elazar, Daniel J., People and Polity: The Organizational Dynamics of World Jewry Wayne State University Press, (1989) ISBN 0-8143-1843-6

Fredman Cernea, Ruth, Almost Englishmen: Baghdadi Jews in British Burma Lexington Books (2007) ISBN 978-0-7391-1647-0

Guang, Pan, China and Israel – 50 years of Bilateral Relations, 1948-98, New York: The Asian and Pacific Rim Institute of the American Jewish Committee, (1999)

Jackson, Stanley, The Sassoons, portrait of a dynasty, London Heinemann (1968)

Katz, Nathan, Who are the Jews of India? The S. Mark Taper Foundation imprint in Jewish studies, University of California Press. (2000) ISBN 978-0-520-21323-4

Lazarus, Baila, The Jews of Khao San Road, Orchid Designs (2004)

Nathan, Eze, The history of Jews in Singapore, 1830-1945, Herbilu Editorial & Marketing Services (1986) ISBN 9971-84-429-X

Needle, Patricia M. (ed.), East Gate of Kaifeng: a Jewish world inside China, China Center, U. of Minnesota, (1992), ISBN 978-0-9631087-0-8

Senor, Dan & Singer, Shaul, Start-up Nation: The Story of Israel's Economic Miracle, Council on Foreign Relations (2009), ISBN 1455502391

Shatzkes Pamela, Kobe: A Japanese haven for Jewish refugees, 1940–1941. Japan Forum, 1469-932X, Volume 3, Issue 2, (1991)

Slapak, Orpa, The Jews of India: A Story of Three Communities. The Israel Museum, Jerusalem (2003) ISBN 965-278-179-7

Steinsaltz, Rabbi Adin, Avot, the Wisdom of our Fathers, The Israel Institute for Talmudic Publications, ISBN-7050040-1940-6/B

Weisz, Tiberiu, The Kaifeng Stone Inscriptions: The Legacy of the Jewish Community in Ancient China, iUniverse New York, (2006) ISBN 0-595-37340-2

Xu Xin, The Jews of Kaifeng, China. History, Culture, and Religion, KTAV Publishing House, Inc, (2003) ISBN 0-88125-791-5

INDEX